ILLUSTRATED KEY TO THE SEAWEEDS OF NEW ENGLAND

Martine Villalard-Bohnsack

A Publication of
The Rhode Island Natural History Survey
Kingston, RI

1995

Published by
The Rhode Island Natural History Survey
Cooperative Extension Education Center
East Alumni Avenue, University of Rhode Island
Kingston, RI 02881-0804

Additional copies may be ordered from the above address.

Library of Congress Catalog Number 95-70423

ISBN 1-887771-00-X (paperbound)
10 9 8 7 6 5 4 3 2
Printed in the United States of America

Financial assistance for the preparation of this book was provided by grants from the Roger Williams University Research Foundation.

A generous grant from the Lamb Foundation supported the publication of this book.

Photos on cover page, from left to right:
Laminaria digitata, Phycodrys rubens, Ulva lactuca

TABLE OF CONTENTS

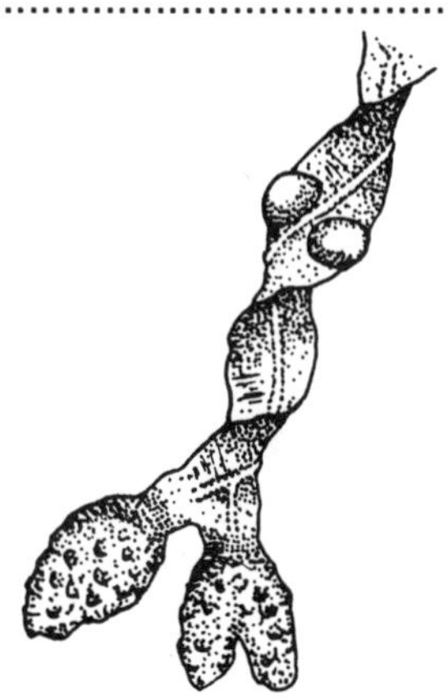

Fucus vesiculosus var. *spiralis*

Dedicated to my parents,
Guy and Madeleine Villalard

Phycodrys rubens

Antithamnion pectinatum

INTRODUCTION

The following guide to the seaweeds of New England was designed to make identification of algal specimens as easy and enjoyable as possible for students of phycology and related fields, or for anyone with an interest in marine biology. A second objective was to produce an affordable alternative to the more technical but expensive treaties. The key includes 206 algal taxa (including species, subspecies and non generic entities) commonly found on the open coast, estuaries and bays of New England. Over 250 species have been recorded in the literature for this area (Bird and McLachlan, 1992; Mathieson and Hehre, 1986; Schneider *et al.*, 1979; Sears and Wilce, 1975; South and Tittley, 1986; Taylor, 1957; Villalard-Bohnsack *et al.*, 1988), but a number of these are uncommon or rare. The selection of species for inclusion in this guide was based on specimens students in my classes have repeatedly collected in New England, over the past twentyfour years, as well as species reported as common by the above authors. Most species included are easily found and can be observed or collected either from the intertidal zone, at low tide, or from the drift, particularly after a storm. Subtidal, epilithic crusts are the exception and are best collected by scuba diving or by dredging.

The key was written for use in an undergraduate science program and does not require an extensive understanding of terminology. Descriptive, common terms are used whenever possible, and when unavoidable, technical terms are defined or explained with illustrations throughout the work. A glossary (p. 134) is also included.

The key was constructed using mostly distinct, macroscopic features and environmental characteristics, instead of systematic criteria. Reproductive structures and cellular details which might be of critical importance to the phycologist, have been kept to a minimum for the purpose of this work, and are only used when necessary. Organisms are grouped in categories such as crusts, flat blades, filaments, etc. Groupings are, therefore, artificial and the keys of the three major divisions do not follow scientific phylogenetic classification.

The nomenclature used for most of the species included was originally based on South and Tittley (1986). However, many changes have been published since then, and most have been incorporated. Recent changes in nomenclature were compiled from a variety of sources, including personal communications from M. Wynne and C. Schneider, as well as works by Bird and McLachlan (1992), Burrows (1991), Fletcher (1987), Irvine and Chamberlain (1994), Maggs and Hommersand (1993), and from original articles. Checklists for the three divisions are organized in orders and families based mostly on South and Tittley (1986) for the Chlorophyta and the Phaeophyta, and on Bird and McLachlan (1992) for the Rhodophyta. Species listed are followed by authorities, along with selected synonyms, to allow for cross reference with earlier keys. Authorities are abreviated and spaced according to Brummitt and Powell (1992). Both detailed synonymy and references on recent nomenclature changes may be found in the references cited above. Exceptions are explained or referenced in the key.

Descriptions of taxa are not meant to be comprehensive and include only enough information for identification purposes. Further information on species morphology, reproduction, life cycles, habitats and distribution can be found in more detailed or more technical works, such as Adey and Adey (1973), Bird and McLachlan (1992), Burrows (1991), Dixon and Irvine (1977), Fletcher (1987), Irvine (1983), Irvine and Chamberlain (1994), Maggs and Hommersand (1993), Schneider and Searles (1991) or Taylor (1957). Information on habitats and distribution was summarized from Adey (1973), Mathieson and Hehre (1986), Sears and Wilce (1975), Taylor (1957), and Villalard-Bohnsack *et al.* (1988). Most descriptions are accompanied by photographs of live specimens or herbarium mounts. In some cases, drawings are used instead of photographs to accentuate diagnostic characters. These were prepared from living organisms, herbarium specimens or from permanent slides. Voucher slides for *Antithamnion cruciatum, A. pectinatum* and *Scagelia pylaisaei* were graciously provided by J. Foertch. Drawings of microscopic features for some species, such as the calcareous and fleshy crusts, and for certain species of *Sphacelaria*, were modified and redrawn from various sources and the appropriate reference is given in the key. The bibliography includes all references used for the preparation of this key, as well as earlier guides to the seaweeds of New England and of adjacent regions.

ACKNOWLEDGEMENTS

I would like to thank the following people for their careful review of the manuscript and for their suggestions: Art Mathieson, Charlie Yarish, Jim Foertch, John Swenarton and Bill Johansen. I am also grateful for the comments and recommendations made by colleagues who "field tested" the key with their respective botany or phycology classes: Marilyn Harlin, Tom Philbrick, Jim Sears, Peter Siver and Charlie Yarish. The members of my own Phycology classes of the spring of 1993-1995 were also involved in the field testing and I am very appreciative of their help. I am also deeply indebted to M. Wynne and C. Schneider for their detailed scrutiny of the nomenclature. The Roger Williams University Research Foundation provided financial assistance for the preparation of the photographs, and I am thankful for that support. Finally, I would like to thank my husband for his enthusiastic encouragements, his patience, his help in proof reading the manuscripts, and for his endless willingness to tackle computer glitches.

Martine Villalard-Bohnsack
Department of Biology
Roger Williams University
Bristol, Rhode Island 02809

USE OF THE KEY

Specimens collected for identification should be in good condition. An ideal specimen includes all vegetative parts and has a true, healthy color. A careful examination of field populations during collecting will ensure that a typical, healthy specimen is chosen. Macroscopic characters can be observed on either fresh or dried and pressed algae; however, microscopic structures such as cellular details are best seen on fresh materials. Many algae can be identified using macroscopic characters only, but certain species require the use of a microscope for observation and for measurements of cells and organelles. Although a good quality hand lens may be sufficient in some cases, a microscope, equipped with an ocular micrometer is necessary at times.

The guide consists of three independent keys, one for each of the three major divisions of seaweeds: Chlorophyta (green algae: 40), Phaeophyta (brown algae: 64), Rhodophyta (red algae: 102). Pigmentation, which is the most visible character selected to classify algae, should be used as the initial criterion for determining the appropriate division (Key to the three major divisions of New England seaweeds, p. 11). However, color, the result of pigmentation, can be extremely variable, and a simple iodine test described in the key, may be used to separate the green algae from the other two divisions. The traditional boiling water test used to separate the red algae from the browns is not reliable and can not be recommended here.

Throughout the key, diagnostic characters are designated in bold letters and in a large print, *e.g.*, **"Thallus a flat blade"**. These are followed by further information, in smaller plain text, to help clarify the broad statements and ensure a correct identification, *e.g.*, "thallus rosy red, up to 50 cm long; found only in spring". This eliminates problems caused by subjective interpretations of terms such as slippery, wiry, etc. A short description of each species, as well as environmental information is included. Thus, specimens can be rapidly keyed out by using the bold characters only. The additional information should be used at the end of the keying process, to verify an identification, or at any time there is a question of interpretation of diagnostic characters. Species that are variable and exhibit different morphological forms are listed at more than one step. The names of sporophytic or gametophytic phases belonging to the life cycles of other genera are used, but placed in quotation marks, and the genus and species of the altemate form are given. When additional species have been recorded for our area but are not common enough to be included in this work, the following notation is used: (♦ 2 of 7), indicating, in this instance, that for this particular genus, only the two most common species out of a possible 7 are included. Therefore, other keys should be consulted when a specimen does not fit any of the enclosed descriptions, since it may be an uncommon or rare species excluded from this work. To this end, additional keys useful for seaweed identification in New England, as well as in a few other selected areas, are listed in the bibliography.

Illustrations are located on the opposite page from the description of a species which is included in last step of the keying process. Drawings are also used throughout the key to clarify or supplement definitions. Photographs and drawings were chosen to best illustrate the characteristics listed. In some cases the photographs have been touched up with pen and ink to emphasize key features. Cellular details are given only when needed for identification and microscopic examination is then necessary. Cross sections are required for some species. These can easily be obtained by chopping a small amount of material to very fine, thread like pieces, using a razor blade on a glass slide. The resulting pieces will then be thin enough to turn over on their side and present a cross view visible with a microscope.

Authorities (names of phycologists who discovered, described and named or renamed the species) and selected synonyms (generic and specific names which are obsolete, but found in earlier keys) are included in the check-lists. Abreviations follow Brummit and Powell (1992). Details on synonymy and authorities may be found in the references listed in the introduction. Complete descriptions may also be found in the above references, as well as in Taylor (1957). Information on uses, history, name derivations and other pertinent facts for some of the species included here, may be found in works such as Kingsbury (1969) and Lee (1986).

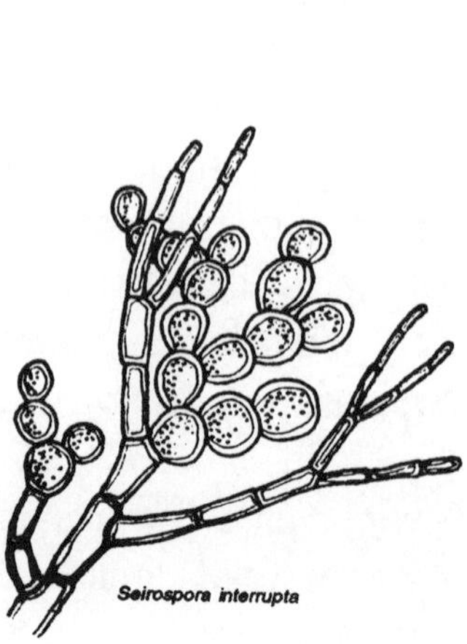

Seirospora interrupta

Checklist of the New England seaweeds included in the following keys (with selected synonyms in parenthesis)

Chlorophyta (40 taxa)

Ulotrichales
Ulotrichaceae
Ulothrix flacca (Dillwyn) Thuret *in* Le Jolis
Ctenocladales
Ctenocladaceae
Pseudendoclonium submarinum Wille (incl. *Protoderma marinum)*
Ulvales
Percursariaceae
Percursaria percursa (C.Agardh) Bory
Ulvaceae
Capsosiphon fulvescens (C.Agardh) Setch. *et* N.L.Gardner
C. groenlandicum (J.Agardh) K.L.Vinogr. *(Enteromorpha groenlandica)*
Enteromorpha clathrata (Roth) Grev.
E. compressa (L.) Nees
E. flexuosa (Wulfen *ex* Roth) J.Agardh ssp. *flexuosa* (Dillwyn) Bliding
E. flexuosa (Wulfen *ex* Roth) J.Agardh ssp. *paradoxa* (Dillwyn) Bliding (incl. *E. erecta, E. plumosa)*
E. intestinalis (L.) Nees
E. linza (L.) J.Agardh
E. prolifera (O.F.Müll.) J.Agardh
Ulva lactuca L.
Ulvaria obscura (Kütz.) Gayral var. *blyttii* (Aresch.) Bliding *(Monostroma fuscum)*
Monostromataceae
Blidingia minima (Nägeli *ex* Kütz.) Kylin *(Enteromorpha minima)*
"*Gomontia polyrhiza* (Lagerh.) Bornet *et* Flahault"
Kornmannia leptoderma (Kjellm.) Bliding *(Monostroma leptodermum)*
Monostroma grevillei (Thuret) Witt. *(Ulvopsis grevillei)*
M. oxyspermum Kütz. *(Ulvaria oxysperma)*
Protomonostroma undulatum (Wittrock) K.L.Vinogr. f. *pulchrum* (Farl.) M.J.Wynne *(Monostroma undulatum,* and incl. *M. pulchrum)*
Prasiolales
Prasiolaceae
Prasiola stipitata Suhr *in* Jessen
Acrosiphoniales
Acrosiphoniaceae
Acrosiphonia arcta (Dillwyn) J.Agardh *(Spongomorpha arcta,* and incl. *A. spinescens = S. spinescens)*
"*Codiolum gregarium* A.Braun"
"*C. petrocelidis* Kuck."
"*C. pusillum* (Lyngb.) Kjellm."
Spongomorpha aeruginosa (L.) C.Hoek *(S. lanosa)*
Urospora penicilliformis (Roth) Aresch.
U. wormskjoldii (Mert. *in* Hornem.) Rosenv. (incl. *U. collabens)*
Cladophorales
Cladophoraceae
Chaetomorpha aerea (Dillwyn) Kütz.
C. brachygona Harv.
C. linum (O.F.Müll.) Kütz.
C. melagonium (Weber *et* D.Mohr) Kütz.
C. picquotiana Mont. *ex* Kütz. *(C. atrovirens)*

Cladophora albida (Nees) Kütz. (incl. *C. glaucescens* of some authors, *C. magdalenae, C. refracta)*
C. pygmaea Reinke
C. rupestris (L.) Kütz
C. sericea (Huds.) Kütz. (incl. *C. flexuosa* f. *densa, C. glaucescens* of some authors, *C. gracilis* of some authors, *C. rudolphiana* of some authors)
Rhizoclonium tortuosum (Dillwyn) Kütz. (incl. *R. riparium)*

Codiales

Bryopdidaceae

Bryopsis plumosa (Huds.) C.Agardh

Codiaceae

Codium fragile (Suringar) Har. ssp. *tomentosoides* (Goor) P.C.Silva

Phaeophyta (64 taxa)

Ectocarpales

Ectocarpaceae

Ectocarpus fasciculatus Harv.
E. siliculosus (Dillwyn) Lyngb. (incl. *E. confervoides, E. dasycarpus, E. penicillatus)*
Hincksia granulosa (J.E.Smith) P.C.Silva *in* P.C.Silva *(Giffordia granulosa)*
H. mitchelliae (Harv.) P.C.Silva *in* P.C.Silva *(Giffordia mitchelliae)*
Laminariocolax tomentosoides (Farl.) Kylin *(Ectocarpus tomentosoides)*
Pilayella littoralis (L.) Kjellm. *(Pylaiella littoralis)*
Spongonema tomentosum (Huds.) Kütz. *(Ectocarpus tomentosus)*

Lithodermataceae

Petroderma maculiforme (Wollny) Kuck. *(Lithoderma maculiforme, L. lignicola)*
Pseudolithoderma extensum (P.Crouan *et* H.Crouan) S.Lund *(Lithoderma extensum)*
P. paradoxum Sears *et* Wilce

Sorocarpaceae

Botrytella micromora Bory *(Sorocarpus micromorus)*

Myrionemataceae

Myrionema corunnae Sauv.
M. strangulans Grev.
Protectocarpus speciosus (Børgesen) Kuck. *(Ectocarpus speciosus)*
Ulonema rhizophorum Fosl.

Elachistaceae

Elachista fucicola (Velley) Aresch. *(Elachistea fucicola)*
Halothrix lumbricalis (Kütz.) Reinke

Chordariales

Leathesiaceae

Leathesia difformis (L.) Aresch.

Spermatochnaceae

Stilophora rhizodes (Turn.) J.Agardh

Acrothrichaceae

Acrothrix gracilis Kylin (incl. *A. novae-angliae)*

Chordariaceae

Chordaria flagelliformis (O.F.Müll.) C.Agardh
Cladosiphon zosterae (J.Agardh) Kylin *(Eudesme zosterae)*
Eudesme virescens (Carmich. *ex* Harv. *in* Hook.) J.Agardh
Sphaerotrichia divaricata (C.Agardh) Kylin

Sphacelariales

Sphacelariaceae

Sphacelaria arctica Harv. *(S. racemosa* var. *arctica)*
S. cirrosa (Roth) C.Agardh

S. plumigera Holmes
S. plumosa Lyngb. *(Chaetopteris plumosa)*
S. racemosa Grev.
S. radicans (Dillwyn) C.Agardh

Cladostephaceae

Cladostephus spongiosus (Huds.) C.Agardh *(C. verticillatus)*

Desmarestiales

Desmaretiaceae

Desmarestia aculeata (L.) J.V.Lamour.
D. viridis (O.F.Müll.) J.V.Lamour.

Arthrocladiaceae

Arthrocladia villosa (Huds.) Duby

Dictyosiphonales

Striariaceae

Isthmoplea sphaerophora (Carmich. *ex* Harv. *in* Hook.) Kjellm.

Punctariaceae

Asperococcus fistulosus (Huds.) Hook. *(A. echinatus)*
Punctaria latifolia Grev.
P. plantaginea (Roth) Grev.
P. tenuissima (C.Agardh) Grev. *(Desmotrichum undulatum)*

Dictyosiphonaceae

Dictyosiphon foeniculaceus (Huds.) Grev.

Scytosiphonaceae

Petalonia fascia (O.F.Müll.) Kuntze
P. zosterifolia (Reinke) Kuntze
Ralfsia fungiformis (Gunn.) Setch. *et* N.L.Gardner
R. verrucosa (Aresch.) J.Agardh
"Stragularia clavata" (Harv. *in* Hook.) Hamel *("Ralfsia bornetii", "R. clavata")*
Scytosiphon simplicissimus (Clemente) Cremades *in* Cremades *et* Pérez-Cirera *(S. lomentaria)*

Laminariales

Chordaceae

Chorda filum (L.) Stackh.
C. tomentosa Lyngb.

Laminariaceae

Agarum clathratum Dumort. *(A. cribrosum)*
Laminaria digitata (Huds.) J.V.Lamour. *(*incl. *L. intermedia)*
L. longicruris Bach.Pyl.
L. saccharina (L.) J.V.Lamour. *(*incl. *L. agardhii)*
Saccorhiza dermatodea (Bach.Pyl.) J.Agardh

Alariaceae

Alaria esculenta (L.) Grev.

Fucales

Fucaceae

Ascophyllum nodosum (L.) Le Jolis
A. nodosum (L.) Le Jolis ecad *scorpioides* (Reinke) Hauck
Fucus distichus L. ssp. *edentatus* (Bach.Pyl.) Powell *(F. edentatus)*
F. distichus L. ssp. *evanescens* (C. Agardh) Powell *(F. evanescens)*
F. spiralis L.
F. vesiculosus L.
F. vesiculosus L. var. *spiralis* Farl.

Sargassaceae

Sargassum filipendula C.Agardh
S. fluitans Børgesen
S. natans (L.) Gaillon

Rhodophyta
(102 taxa)

Porphyridiales
Porphyridiaceae
Stylonema alsidii (Zanardini) K.M.Drew *(Goniotrichum alsidii)*
Compsopogonales
Erythropeltidaceae
Erythrotrichia carnea (Dillwyn) J.Agardh
Bangiales
Bangiaceae
Bangia atropurpurea (Roth) C.Agardh *(*incl. *B. fuscopurpurea)*
Porphyra leucosticta Thur. *in* Le Jolis
P. linearis Grev.
P. miniata (C.Agardh) C.Agardh
P. umbilicalis (L.) J.Agardh
Acrochaetiales
Acrochaetiaceae
Audouinella spp. *(Acrochaetium* sp., *Rhodochorton* sp., or *Kylinia* sp., depending on species*)*
Nemaliales
Liagoraceae (Helminthocladiaceae)
Nemalion helminthoides (Velley *in* With.) Batters *(*incl. *N. multifidum)*
Bonnemaisoniales
Bonnemaisoniaceae
Bonnemaisonia hamifera Har. *(Asparagopsis hamifera)*
"*Trailliella intricata* Batters"
Gelidiales
Gelidiaceae
Gelidium pusillum (Stackh.) Le Jolis (incl. *G. crinale*)
Palmariales
Palmariaceae
Devaleraea ramentacea (L.) Guiry *(Halosaccion ramentaceum)*
Palmaria palmata (L.) Kuntze *(Rhodymenia palmata)*
Rhodophysemataceae
Rhodophysema elegans (P.Crouan *et* H.Crouan *ex* J.Agardh) P.S.Dixon *(Rhododermis elegans)*
R. georgii Batt. *(Rhododermis georgii)*
Hildenbrandiales
Hildenbrandiaceae
Hildenbrandia rubra (Sommerf.) Menegh. *(H. prototypus)*
Corallinales
Corallinaceae
Clathromorphum circumscriptum (Strömf.) Fosl. *(Phymatolithon evanescens)*
Corallina officinalis L.
Hydrolithon farinosum (J.V.Lamour.) Penrose *et* Y.M.Chamb. *(Fosliella farinosa , Melobesia farinosa)*
Leptophytum foecundum (Kjellm.) Adey *(Lithothamnion foecundum)*
L. laeve (Strömf.) Adey *(Liththamnion laeve)*
Lithothamnion glaciale Kjellm.
Melobesia membranacea (Esper) J.V.Lamour. *(Epilithon membranaceum)*
Phymatolithon laevigatum (Foslie) Foslie *(Lithothamnion laevigatum)*
P. lamii (Me.Lemoine) Chamb.
P. lenormandii (Aresch. *in* J.Agardh) Adey *(Lithothamnion lenormandii)*
Pneophyllum fragile Kütz. *(P. lejolisii, Fosliella lejolisii, Melobesia lejolisii)*
Titanoderma pustulatum (J.V.Lamour.) Woelk., Y.M.Chamb. et P.C.Silva *(Dermatolithon pustulatum, Lithophyllum pustulatum, L. macrocarpum)*
T. corallinae (P.Crouan *et* H.Crouan) Woelk.

Gigartinales

Choreocolacaceae

Choreocolax polysiphoniae Reinsch

Harveyella mirabilis (Reinsch) F.Schmitz *et* Reinke

Cystocloniaceae

Cystoclonium purpureum (Huds.) Batters var. *cirrhosum* Harv.

Fimbrifolium dichotomum (Lepechkin) G.I.Hansen *(Rhodophyllis dichotoma)*

Dumontiaceae

Dumontia contorta (S.G.Gmelin) Rupr. *(D. incrassata)*

Gigartinaceae

Chondrus crispus Stackh.

Gloiosiphoniaceae

"Cruoriopsis spp."

Gloiosiphonia capillaris (Huds.) Carmich. *ex* Berk.

Halymeniaceae

Grateloupia doryphora (Mont.) Howe

Hypneaceae

Hypnea musciformis (Wulfen *in* Jacqu.) J.V.Lamour.

Kallymeniaceae

Callocolax neglectus F.Schmitz *ex* Batters

Callophyllis cristata (C.Agardh) Kütz. *(Euthora cristata)*

Petrocelidaceae

Mastocarpus stellatus (Stackh. *in* With.) Guiry *in* Guiry et al. *(Gigartina stellata)*

"Petrocelis cruenta J.Agardh" *("P. middendorfii")*

Peyssonneliaceae

Peyssonnelia dubyi P.Crouan *et* H.Crouan

P. rosenvingii F.Schmitz *in* Rosenv.

Phyllophoraceae

Ceratocolax hartzii Rosenv.

Coccotylus truncatus (Pallas) M.J.Wynne *et* Heine *(Phyllophora tuncata, P. brodiaei)*

Gymnogongrus crenulatus (Turn.) J.Agardh *(G. norvegicus)*

G. griffithsiae (Turn.) Martius

Phyllophora pseudoceranoides (Gmelin) Newr. *et* A.Taylor *(P. membranifolia)*

Polyideaceae

Polyides rotundus (Huds.) Grev. *(P. caprinus)*

Solieriaceae

Agardhiella subulata (C.Agardh) Kraft *et* M.J.Wynne *(A. tenera, Neoagardhiella baileyi)*

Turnerella pennyi (Harv.) F.Schmitz

Ahnfeltiales

Ahnfeltiaceae

Ahnfeltia plicata (Huds.) Fries

Gracilariales

Gracilariaceae

Gracilaria tikvahiae McLachlan *(*Incl. *G. foliifera var. angustissima, G. verrucosa* of some authors*)*

Rhodymeniales

Champiaceae

Champia parvula (C.Agardh) Harv.

Lomentariaceae

Lomentaria baileyana (Harv.) Farl.

L. clavellosa (Turn.) Gaillon

L. orcadensis (Harv.) Collins *ex* W.R.Taylor

Ceramiales

Ceramiaceae

Aglaothamnion roseum (Roth) Maggs *et* L'Hardy-Halos *(Callithamnion roseum)*

A. westbrookiae Rueness *et* L'Hardy-Halos *(Callithamnion byssoides)*

Antithamnion cruciatum (C.Agardh) Nägeli

A. pectinatum (Mont.) Brauner *in* Athanasiadis *et* Tittley *(Callithamnion pectinatum)*
Antithamnionella floccosa (O.F.Müll.) Whittick *(Antithamnion floccosum)*
Callithamnion corymbosum (J.E.Smith) Lyngb.
C. tetragonum (With.) S. F.Gray (incl. *C. baileyi)*
Ceramium cimbricum H.E.Petersen *in* Rosenv. *(C. fastigiatum, C. fastigiaramosum)*
C. deslongchampii Chauv. *ex* Duby
C. diaphanum (Lightf.) Roth
C. nodulosum (Lightf.) Ducluzeau *(C. rubrum*, incl. *C. areschougii, C. pedicellatum, C. rubriforme)*
C. strictum Harv.
Griffithsia globulifera Harv. *ex* Kütz. *(G. globifera)*
Pleonosporium borreri (Sm.) Nägeli *ex* Hauck
Plumaria plumosa (Huds.) Kuntze *(P. elegans)*
Pterothamnion plumula (J.Ellis) Nägeli *(Antithamnion plumula)*
Ptilota serrata Kütz.
Scagelia pylaisaei (Mont.) M.J.Wynne *(S. corallina, Antithamnion americanum, A. pylaisaei)*
Seirospora interrupta (J.E.Smith) F.Schmitz *(S. seirosperma, S. griffithsiana)*
Spermothamnion repens (Dillwyn) Rosenv. *(S. turneri)*
Spyridia filamentosa (Wulfen) Harv. *in* Hook.

Delesseriaceae

Calloglossa leprieurii (Mont.) J.Agardh
Grinnellia americana (C.Agardh) Harv.
Membranoptera alata (Huds.) Stackh. (incl. *M. denticulata)*
Pantoneura baerii (Post. *et* Rupr.) Kylin
Phycodrys rubens (L.) Batters

Dasyaceae

Dasya baillouviana (S.G.Gmelin) Mont. *(D. pedicellata)*

Rhodomelaceae

Bostrychia radicans (Mont.) Mont.
Chondria baileyana (Mont.) Harv.
C. capillaris (Huds.) M.J.Wynne *(C. tenuissima)*
C. dasyphylla (Woodw.) C.Agardh
Polysiphonia denudata (Dillwyn) Grev. *ex* Harv. *in* Hook.
P. elongata (Huds.) Spreng.
P. fibrillosa (Dillwyn) Spreng.
P. flexicaulis (Harv.) Collins
P. fucoides (Huds.) Grev. (incl. *P. nigrescens)*
P. harveyi Bailey (incl. *P. novae-angliae)*
P. lanosa (L.) Tandy
P. nigra (Huds.) Batters
P. stricta (Dillwyn) Grev. (*P. urceolata*)
P. subtilissima Mont.
Rhodomela confervoides (Huds.) P.C.Silva

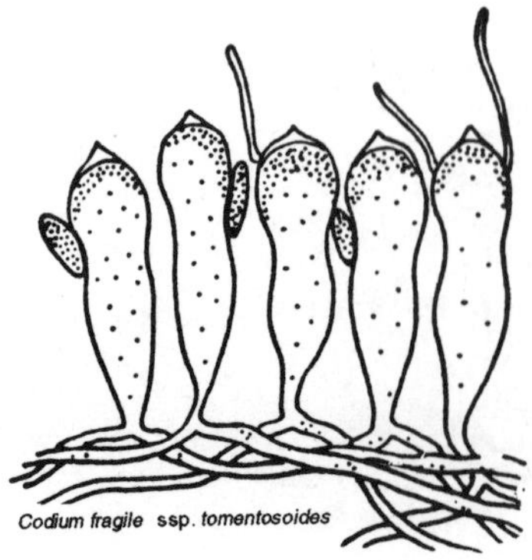

Codium fragile ssp. *tomentosoides*

KEY TO THE THREE MAJOR DIVISIONS OF NEW ENGLAND SEAWEEDS

1. **Thallus generally grass green in color**, but ranging from pale yellowish green, to light or dark olive-green, to brownish or black; **staining purple or blue-black with iodine*** .... **Chlorophyta** (Green algae); p. 13

1. **Thallus generally brown, pink, red, or purplish in color**, but ranging from yellowish or greenish, to olive-brown or black; **not staining purple or blue-black with iodine* (iodine remains brownish)** 2

2. **Thallus light to dark brown,** but sometimes black, greenish brown or yellowish .... **Phaeophyta** (Brown algae); p. 37

2. **Thallus usually pale to rosy pink, light to dark red, or purplish,** but sometimes straw color, greenish or black .... **Rhodophyta** (Red algae); p. 71

* A solution of potassium iodide (I_2KI) can easily be made by mixing the following: 2 g KI, 1 g I, and 25 ml water. The true starch of the Chlorophyta will turn a purple to blue-black color when placed in contact with this solution. A very small drop added to a microscope slide preparation will also highlight Chlorophyta pyrenoids. Information on pyrenoid numbers is often used to identify species within this division. In this case, care must be taken to add only the smallest possible drop, since an excess of I_2KI will turn most Chlorophyta material so dark, that cellular details will be obscured.

Codium fragile

KEY TO THE CHLOROPHYTA

Major diagnostic characters

14 (Green algae)

1. **Green crusts, slicks or stains on rocks, shells or wood** 2
1. **Thallus not crustose or stain-like** 4

2. **Very thin crust, forming a green stain on old shells, occasionally on *Spirorbis* sp. and on barnacles;** made up of creeping, branched filaments, 4-8 µm in diameter; filaments septate, penetrating into the shell; often with swollen sac-like sporangia; found year round, on the open coast, throughout the intertidal zone and just below. The status of *Gomontia* is uncertain. It may be a stage in the life history of *Monostroma grevillei* and may no longer be a valid taxon. In the British Isles, as many as six different taxa have been confused under this name (Wilkinson and Burrows, 1972). Studies involving culture work of the New England complex are required to clarify its status .... "*Gomontia polyrhiza*" (Fig. 1)
2. **Thallus crust-like or a soft slick, thicker than a stain** 3

3. **Irregular crust, loosely organized; attached to rocks or shells;** branched filaments joined laterally and forming a crust with 3 to 8-celled, upright filaments; 1 pyrenoid per cell; found year round, on the open ocean and in estuaries, from the high intertidal into the subtidal zone, down to -10 m, attached to rocks, shells and wood .... *Pseudendoclonium submarinum* (Fig. 2)
3. **Single, elongated cell, up to 3 mm long, oval at top, tapered at base** (somewhat like a baseball bat); net-like chloroplast with many pyrenoids; found on the open coast and in estuaries, growing year round, but more common in summer and fall. "*C. gregarium*" and "*C. pusillum*" form large, dark green, velvety patches on rocks in the high intertidal zone, often mixed with other algae; "*C. petrocelidis*" is found mostly north of Cape Cod, on or within crusts such as *Ralfsia* sp. and *Petrocelis* sp. *Codiolum* spp. represent the sporophyte generations of *Urospora wormskjoldii, Spongomorpha* spp. and *Acrosiphonia* spp. and are no longer considered valid taxa ... . "*Codiolum*" spp. (Fig. 3)

4. **Thallus filamentous (hair-like) down to the base, uniseriate (one row of cells) or biseriate (two rows of cells),** with or without branches 5

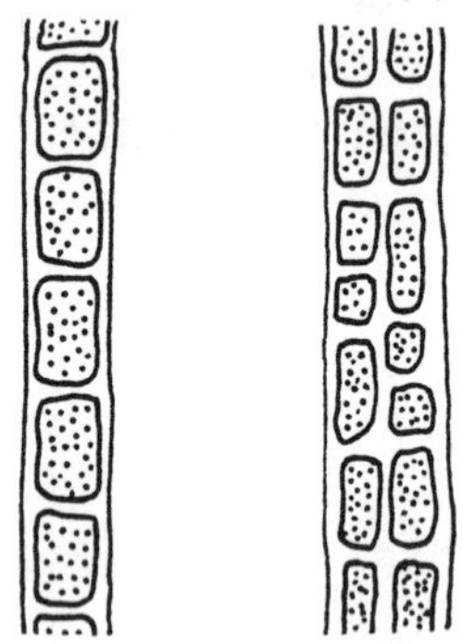

uniseriate filament — biseriate filament

4. **Thallus a blade (flat), or a tube (at least near base);** tube may be solid or hollow, branched or unbranched, and may become filamentous (hair-like) in upper parts 22

5. **Filaments clearly branched** 6
5. **Filaments not branched, or with only a few rare, short, rhizoidal branchlets** 12

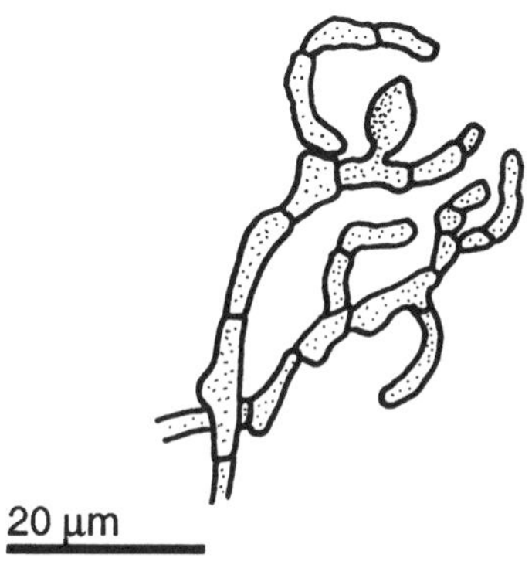

Fig. 1. ***"Gomontia polyrhiza"***
irregularly branched filaments
with a young, swollen sporangium

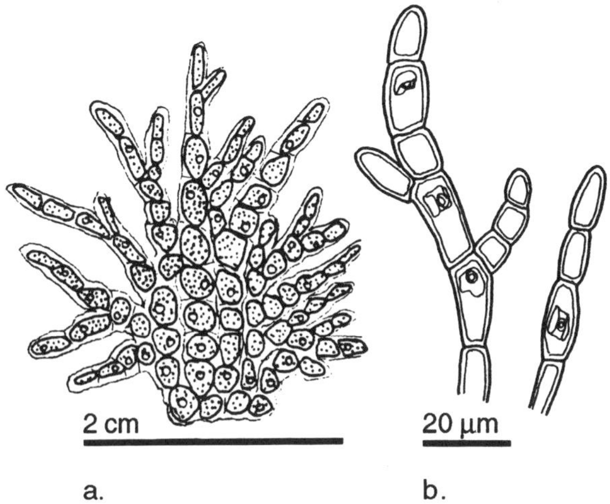

Fig. 2. ***Pseudendoclonium submarinum***
a. surface view of individual crust showing 1 pyrenoid per cell; b. upright filament of 3-8 cells

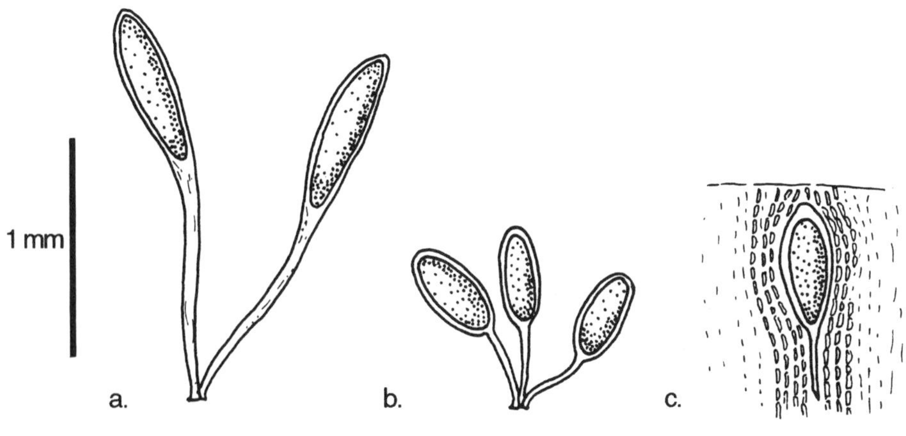

Fig. 3. "***Codiolum***" spp.
a. "*C. gregarium*"; b. "*C. pusillum*";
c. "*C. petrocelidis*" within crustose host

6. **Filaments non-septate** (transversal cell walls absent or rare); main axes with branchlets arranged in a somewhat flattened, triangular or feather-like pattern, branches usually constricted at base; growing singly or in tufts, up to 10 cm long; found year road, but more common spring through fall, on the open coast and in estuaries, in the subtidal zone, down to -10 m, attached to rocks, shells or wood, in somewhat protected places; also found in tide pools; (♦1 of 2) .... *Bryopsis plumosa* (Fig. 4)

6. **Filaments with distinct transversal cell walls** (cells square, rectangular or rounded) 7

7. **Tufts no more than 2 mm high,** dark green, stiff, irregularly branched; found year round, on the open coast and in estuaries, Cape Cod and north, attached to stones and rocks, in the subtidal zone, down to -10 m; this species is difficult to find because of its size; (*Cladophora* spp.♦4 of 7 or more). Specialized keys should be consulted if a specimen does not fit the 4 species of *Cladophora* described here .... *Cladophora pygmaea* (Fig. 5)

7. **Tufts longer than 2 mm** 8

8. **Filaments free, sometimes entangled or slightly twisted, but without specialized branches**; irregular tufts up to 40 cm long 9

8. **Filaments forming rope-like, entangled strands, at least near base, held together by descending rhizoids or recurved branches**; up to 10 cm long; found in late winter to early spring only 11

9. **Branches often opposite or in whorls of 4, strait, dense and broom-like;** tufts dark green, up to 20 cm long, with distinct axes; found year round, on the open coast and in estuaries, north of Cape Cod where it is common, in tide pools and in shaded intertidal and shallow subtidal areas, down to -1 m, attached to rocks .... *Cladophora rupestris* (Fig. 6)

9. **Branching mostly alternate or irregular** 10

17 (Green algae)

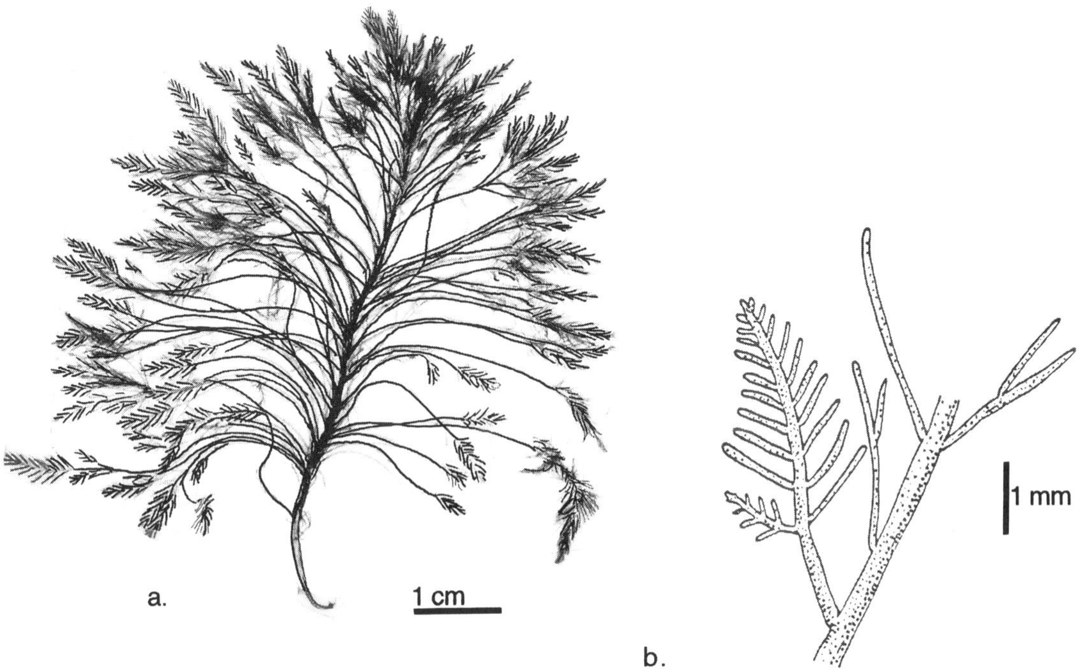

Fig. 4. ***Bryopsis plumosa***
a. whole thallus; b. detail of branchlet showing triangular branching pattern

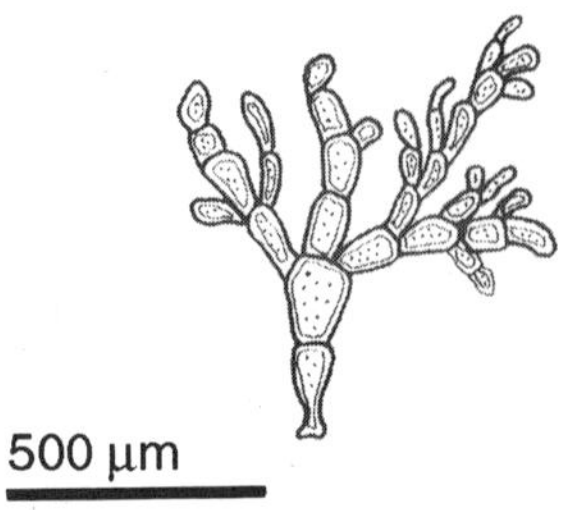

Fig. 5. ***Cladophora pygmaea***
tuft showing irregular branching

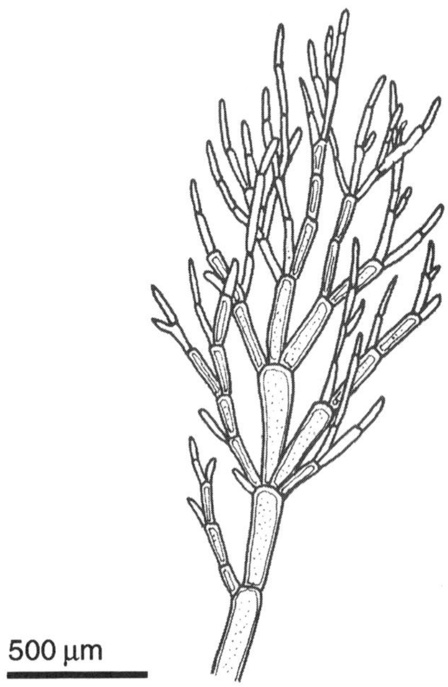

Fig. 6. ***Cladophora rupestris***
detail showing opposite and whorled branches

10. **Branches relatively straight, often in comb-like series** (drawing below); tuft with a rough texture, up to 40 cm long; common year round, on the open coast, and in estuaries and bays, in tide pools throughout the intertidal zone and in the subtidal zone, down to -6 m, attached to rocks or other substrata .... *Cladophora sericea* (Fig. 7)

10. **Branches usually distinctly curved** (drawing below); tufts soft, up to 25 cm long; found year round, on the open coast, in estuaries and in bays growing in tide pools and in the subtidal zone, down to -3 m, attached to rocks and epiphytic on other algae..*Cladophora albida* (Fig. 8)

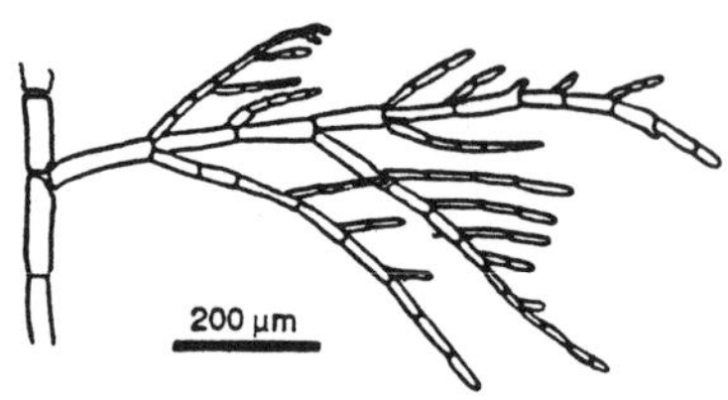

C. sericea , branching pattern

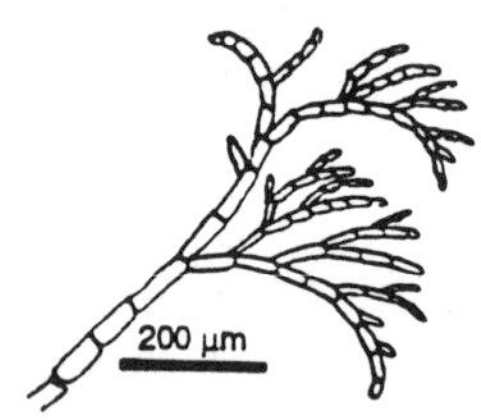

C. albida, branching pattern

11. **Filaments very straight and somewhat stiff; 40-120 μm in diameter, becoming wider towards the tips; growing in rounded, green tufts, up to 10 cm long;** base of tufts often tangled into irregular, rope-like strands; cells 2-20 diameters long; numerous descending rhizoids present in lower part of tuft; spine-like or hooked lateral branches found in older specimens*; common in late winter and spring south of Cape Cod, but found year round to the north, in exposed areas on the open coast, in the low intertidal zone and in the subtidal zone, down to -11 m, attached to rocks; also common in tide pools .... *Acrosiphonia arcta* ** (Fig. 9)

11. **Filaments softer and curved; 30-35 μm in diameter, not widening towards tips; forming green, loose, puff ball-like tufts, up to 5 cm tall;** branches tangled into distinct rope-like strands; cells 2-3 diameters long; found in late winter and spring, in the low intertidal zone; usually epiphytic when young, but often free floating when mature; also found in tide pools .... *Spongomorpha aeruginosa*

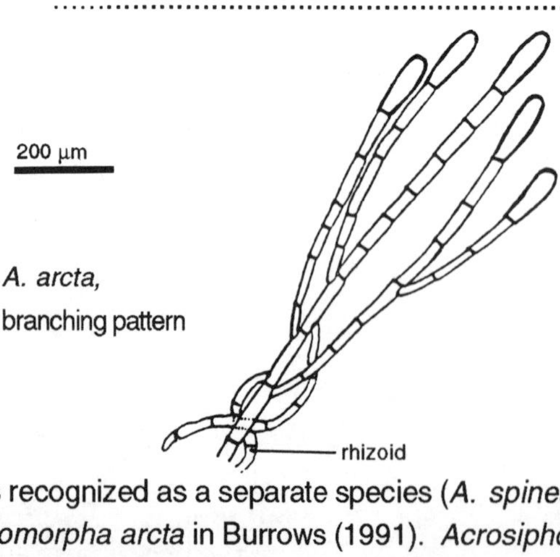

A. arcta, branching pattern

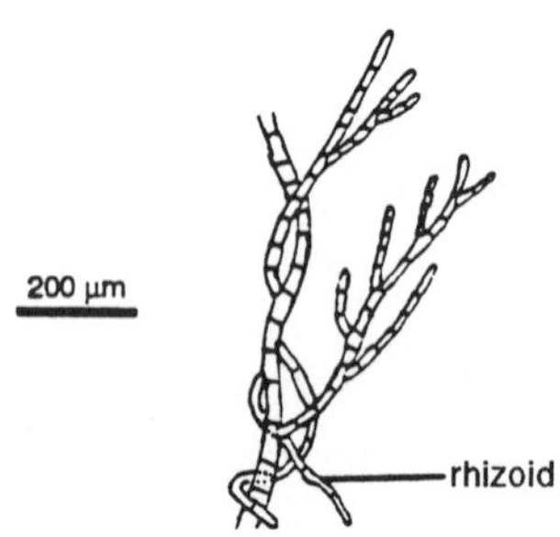

S. aeruginosa, branching pattern

* This form is recognized as a separate species (*A. spinescens*) by some authors (South and Tittley,1986)

** As *Spongomorpha arcta* in Burrows (1991). *Acrosiphonia* and *Spongomorpha* were originally separated on the basis of the nuclear condition (multinuclear vs uninuclear)of the cells(Jónsson, 1962), and that distinction is maintained here.

Fig. 7. ***Cladophora sericea***
whole thallus (tuft)

Fig. 8. ***Cladophora albida***
whole thallus (tuft)

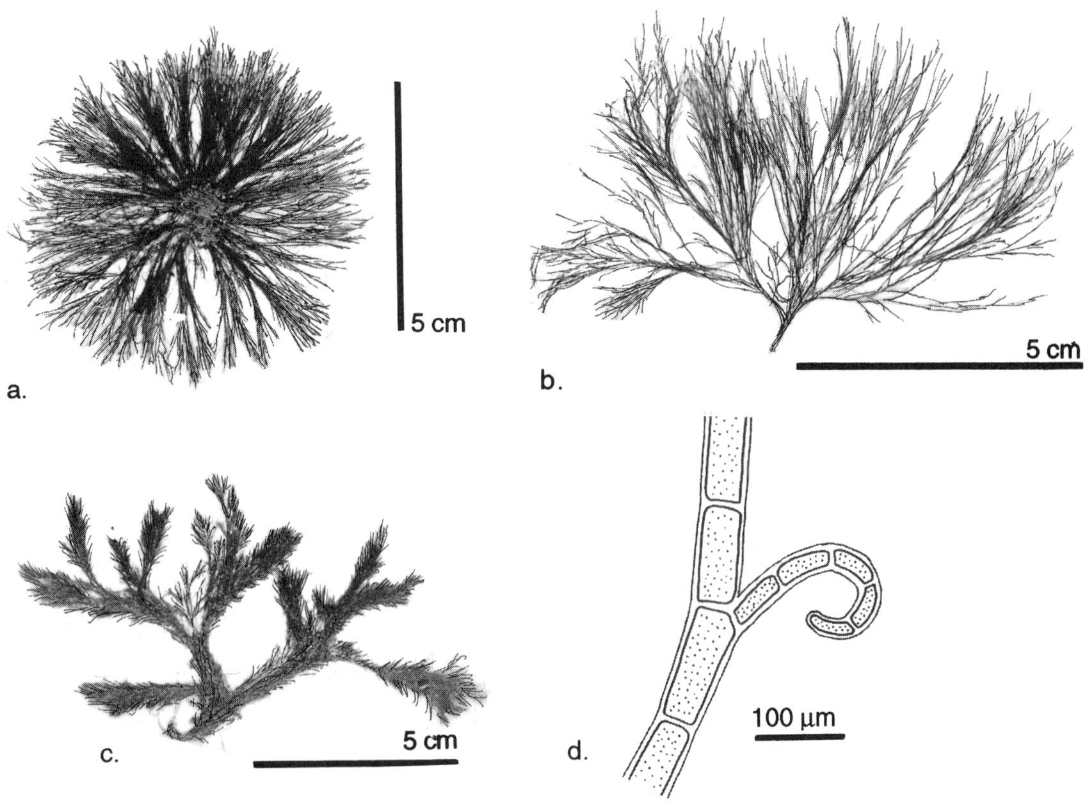

Fig. 9. ***Acrosiphonia arcta***
a. whole thallus (tuft); b. detail of branch; c. rope-like effect on old specimen; d. detail of hooked-shaped lateral branch

12. **Filaments usually biseriate (1-4), delicate; very pale green**; 20-30 μm wide by up to 5 cm long; rarely branched; cells rectangular; found year round, but more common in summer; growing on mud, in salt marshes and estuaries, often forming floating entangled masses with other algae .... *Percursaria percursa* (Fig. 10)
12. **Filaments uniseriate** 13

13. **Chloroplast in the shape of an open bracelet;** filament usually short, entangled, very thin and slippery; cells shorter than wide, cylindrical, 10-25 μm in diameter; cells swollen, and filaments contorted, when reproductive; attached by a single rhizodal cell; abundant in winter and spring, on the open coast and in estuaries, in the intertidal zone, attached to rocks, wood and shells, or epiphytic on other algae; particularly common on fucoids and on flowering plants; (♦1 of 2) .... *Ulothrix flacca* (Fig.11)
13. **Chloroplast otherwise** 14

14. **Regular, transversal cell walls absent, but occasional septa present; tangled masses with horizontal and vertical filaments; forming creeping, shapeless carpet-like mats on any substrate**; found year round; (for adult form see, p. 24-25, step 22, Fig. 21) .... *Codium fragile* ssp. *tomentosoides* (juvenile stage) (Fig. 12)
14. **Transversal cell walls clearly visible with the microscope;** distinct cells present (usually round, square or rectangular) 15

15. **Filaments straight, attached by basal cell** 16
15. **Filaments curly or at least flexuous; often free floating or entangled with other algae** 19

16. **Filaments soft; with long rhizoids produced by several cells above the base** 17
16. **Filaments coarse and stiff, not adhering well to paper when dried; with short rhizoids produced by the basal cell only** 18

17. **Filaments soft and slippery, up to 8 cm long; 30 to 60 μm in diameter; adhering to paper when dried; cells not visible with the naked eye;** vegetative cells cylindrical, but very swollen (barrel shaped) when reproductive; net-like chloroplast; forming dark patches on rocks or wood; found year round, but most common in winter and spring, on the open coast and in estuaries, in the intertidal zone and just below....................... *Urospora penicilliformis* (Fig. 13)
17. **Filament a chain of large, rounded, swollen cells visible with the naked eye; cells 100-500 μm wide, but often 1 mm at tip;** up to 20 cm long, tapering to base; found mostly in late winter and spring, in the intertidal zone, forming slippery patches on rocks and other substrata, in exposed areas on the open coast; often mixed with other algae .... *Urospora wormskjoldii* (Fig. 14)

Fig. 10. ***Percursaria percursa***
detail of biseriate filament showing cells with chloroplasts and pyrenoids

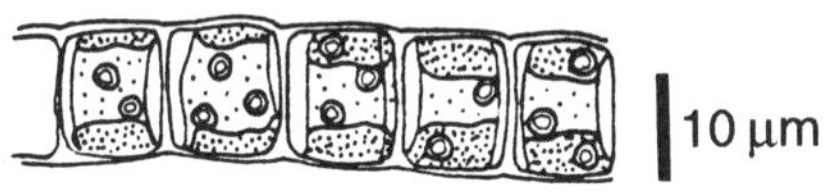

Fig. 11. ***Ulothrix flacca***
detail of filament with chloroplast in the shape of an open bracelet

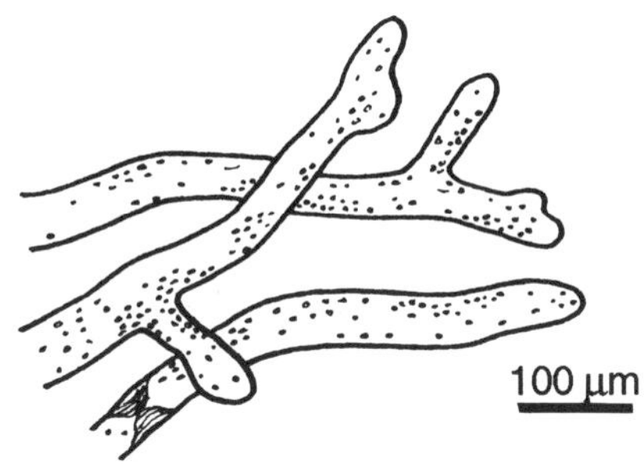

Fig. 12. ***Codium fragile* ssp. *tomentosoides***
juvenile stage, detail of filaments with chloroplasts and septum

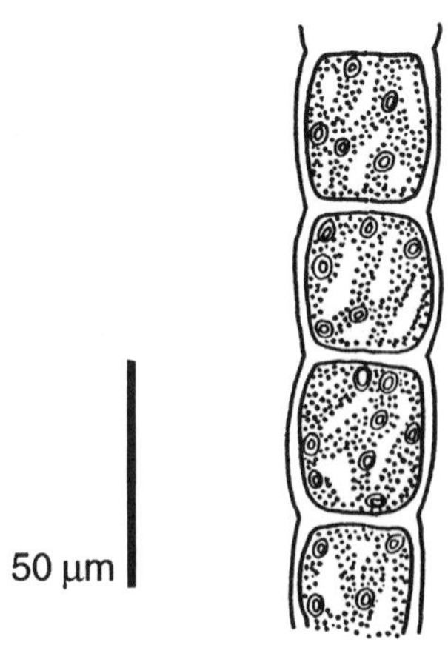

Fig. 13. ***Urospora penicilliformis***
detail of filament with net-like chloroplast

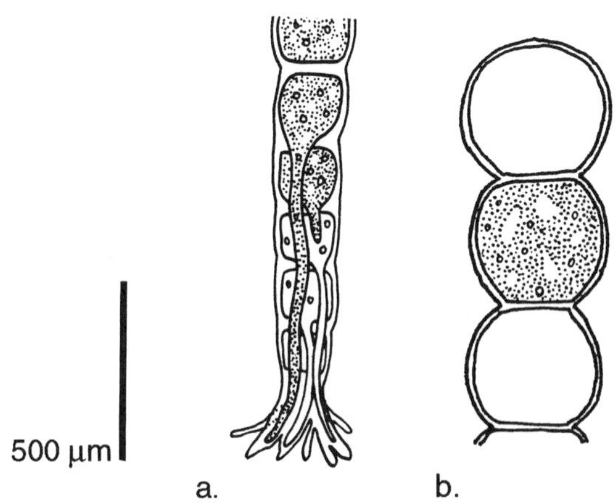

Fig. 14. ***Urospora wormskjoldii***
a. lower end of filament with basal cell and rhizoids; b. tip of filament with large, swollen cells

18. **Cells visible with the naked eye, 300-850 μm wide, rectangular or slightly swollen, 1.5-3 diameters long; basal cell longer (up to 14 diameters long); found in the low intertidal and upper subtidal zones**; filaments straight and stiff, dark green, up to 40 cm long; found singly or in tufts, year round, mostly on the open coast, in the low intertidal and subtidal zones, down to -28 m, attached to rocks or epiphytic on other algae; uncommon south of Cape Cod .... *Chaetomorpha melagonium* (Fig. 15)

18. **Cells 150-350 μm wide, often swollen, 1-2 diameters long or shorter; basal cell longer (from 3-8 diameters long); found in the mid to high intertidal zone;** filaments strait and stiff, bright green, up to 25 cm long, often in tufts or patches; found year round, on the open coast, mostly attached to rocks, in tide pools .... *Chaetomorpha aerea* (Fig. 16)

19. **Filaments soft, cells less than 150 μm in diameter** 20

19. **Filaments coarse, cells greater than 150 μm in diameter;** filaments do not collapse out of water and do not adhere well to paper when pressed and dried 21

20. **Cells 15-70 μm in diameter;** soft, thin, entangled filaments,with an occasional short rhizoidal branch; collapses out of water; adheres well to paper; found year round, on the open coast, in the intertidal zone and just below, attached to rocks; also in tide pools or forming floating mats in estuaries, bays and salt marshes .... *Rhizoclonium tortuosum** (Fig. 17)

20. **Cells 75-150 μm in diameter;** filaments soft, entangled; found year round, on the open coast and in estuaries, in the low intertidal and subtidal zones, down to -18 m; often floating and entangled with other algae .... *Chaetomorpha brachygona* (Fig. 18)

21. **Cells 1-2 diameters long, 150-400 μm wide;** light green, coarse, usually entangled filaments; found year round, on the open coast and in estuaries, in the intertidal zone and below; often forming large free floating masses; most common of the 6 local species .... *Chaetomorpha linum* ** (Fig. 19)

21. **Cells 2-5 diameters long, 200-400 μm wide, cylindrical to barrel-shaped;** dark green, entangled filaments; found year round, on the open coast and in estuaries, in the low intertidal and subtidal zones, down to -20 m, attached to rocks, or entangled with other algae; uncommon south of Cape Cod .... *Chaetomorpha picquotiana* ** (Fig. 20)

* *R. riparium,* included here in *R. tortuosum,* is recognized by some authors as a separate species (Blair, 1983). Others (South and Tittley, 1986), include *R. tortuosum* in *R. riparium.* Usage here, follows that of Burrows (1991).

** *Chaetomorpha linum* and *C. picquotiana* may be detached ecads (ecological variants) of *C. aerea* and *C. melagonium* respectively (Blair, 1983; Blair *et al.*, 1982; Burrows, 1991).

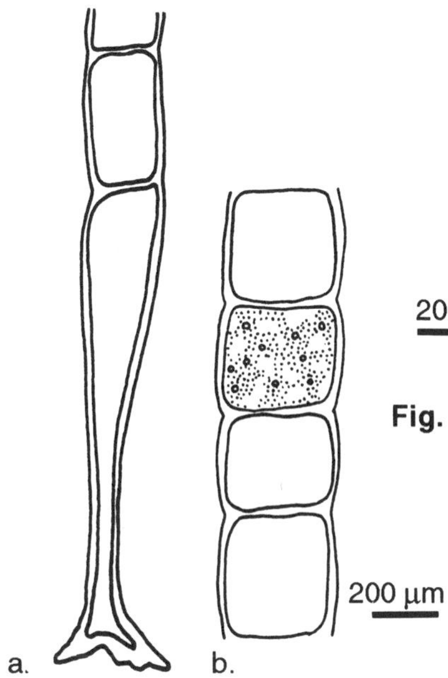

Fig. 15. ***Chaetomorpha melagonium***
a. basal part of filament
b. upper portion of filament with chloroplast and pyrenoids

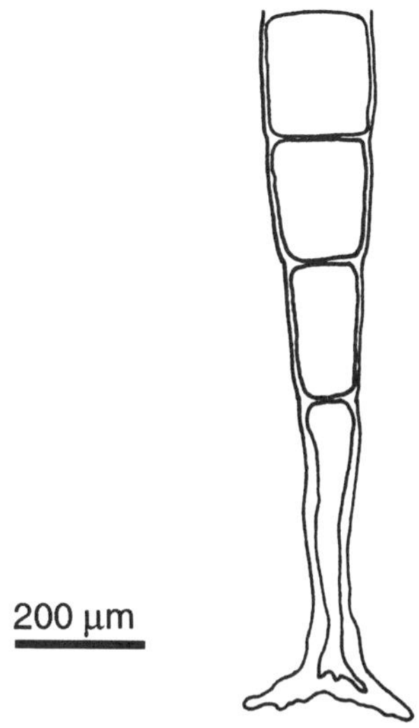

Fig. 16. ***Chaetomorpha aerea***
lower end of filament with basal cell

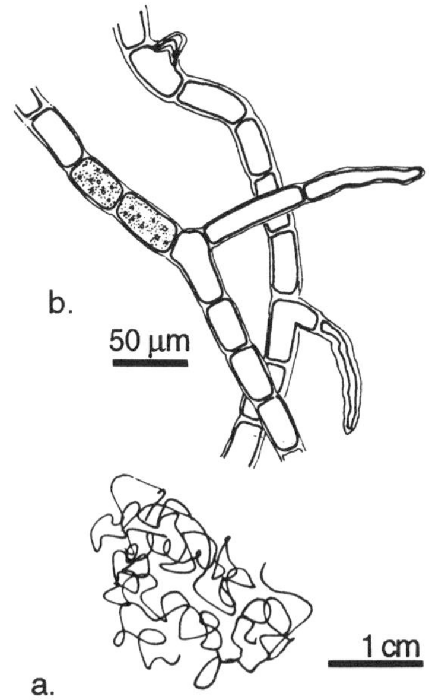

Fig. 17. ***Rhizoclonium tortuosum***
a. whole filament as seen with the naked eye; b. detail of filament with rhizoid

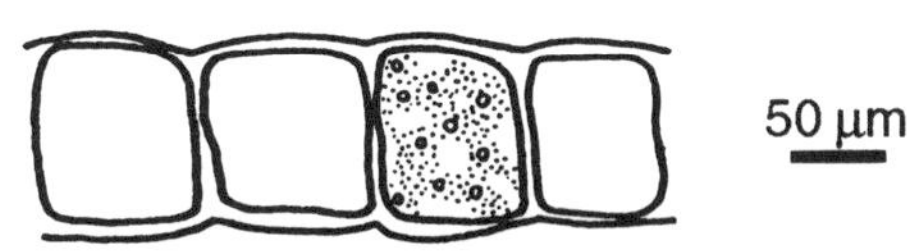

Fig. 18. ***Chaetomorpha brachygona***
detail of filament with chloroplast and pyrenoids

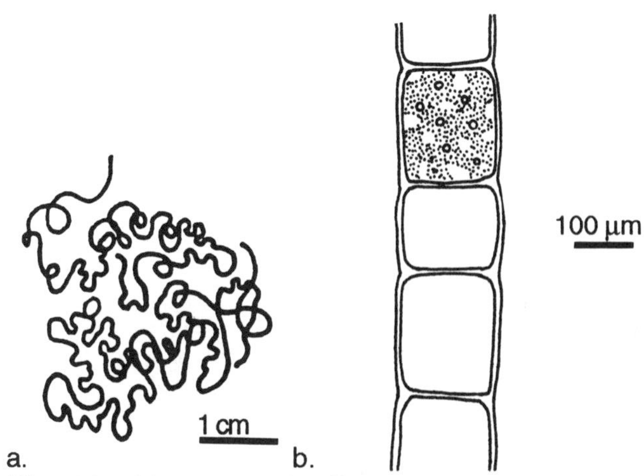

Fig. 19. ***Chaetomorpha linum***
a. whole filament as seen with the naked eye; b. detail of filament with chloroplast and pyrenoids

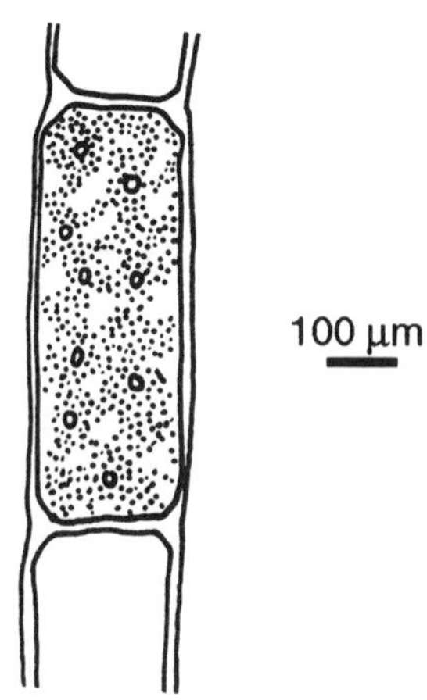

Fig. 20. ***Chaetomorpha picquotiana***
detail of filament with chloroplast and pyrenoids

22. **Large, spongy, cylindrical tubes; dichotomously branched** (even, Y-shaped branches); up to 70 cm long; made up of interwoven filaments ending in a layer of swollen utricles (Fig. 21b) that cover the surface of the thallus; very common year round, on the open coast and in estuaries, in the intertidal and subtidal zones, down to -12 m, attached to rocks, shells and other algae; also found in tide pools; often covered with other epiphytic species such as *Ceramium nodulosum*; (for juvenile stage see, pp. 20-21, step 14, Fig. 12) .... *Codium fragile* ssp. *tomentosoides* (adult form) (Fig. 21)

22. **Thallus otherwise** 23

23. **Thallus a narrow tube (a few micrometers to 5 cm wide), hollow or flattened** 24

23. **Thallus a flat blade, sometimes sac-like when young, but later splitting open into 1 or more flat segments** 33

24. **Unbranched tube, flattened in center, with hollow, undulated edges, resulting in a cross-section typical of this species** (drawing below); tubular at base; up to 40 cm long by up to 5 cm wide; grass green to yellowish green; very common year round, on the open coast and in estuaries, growing singly, in tufts or large patches, in the intertidal and subtidal zones, down to -12 m, attached to rocks, wood and shells, or epiphytic on other algae; (♦1 of 8 or more) .... *Enteromorpha linza* (Fig. 22)

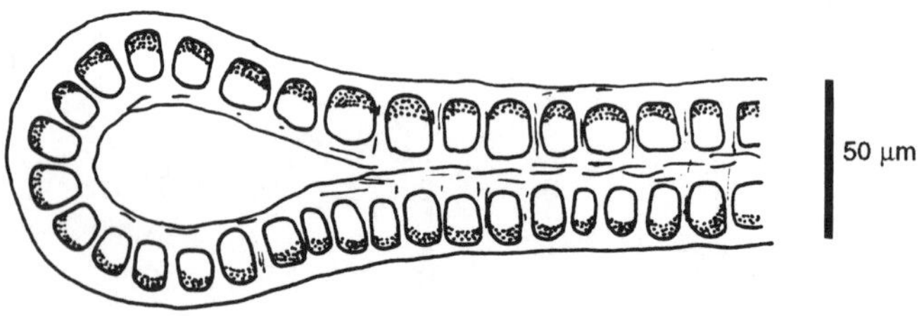

E. linza, portion of cross-section showing hollow margin

24. **Tube hollow, with or without branches** (if present, branches may become filamentous in upper parts of thallus) 25

25. **Thallus rarely branched;** cellular characters are needed for further identification 26

25. **Thallus branched;** cellular characters are needed for further identification 29

26. **Thallus a thin tube, almost hair-like, 5 mm or less in diameter; cells in groups of 2 or 4, at least near base** 27

26. **Thallus otherwise; cells irregularly arranged;** (♦*Blidingia* spp. 1 of 2; ♦*Enteromorpha* spp. 6 of 8 or more) 28

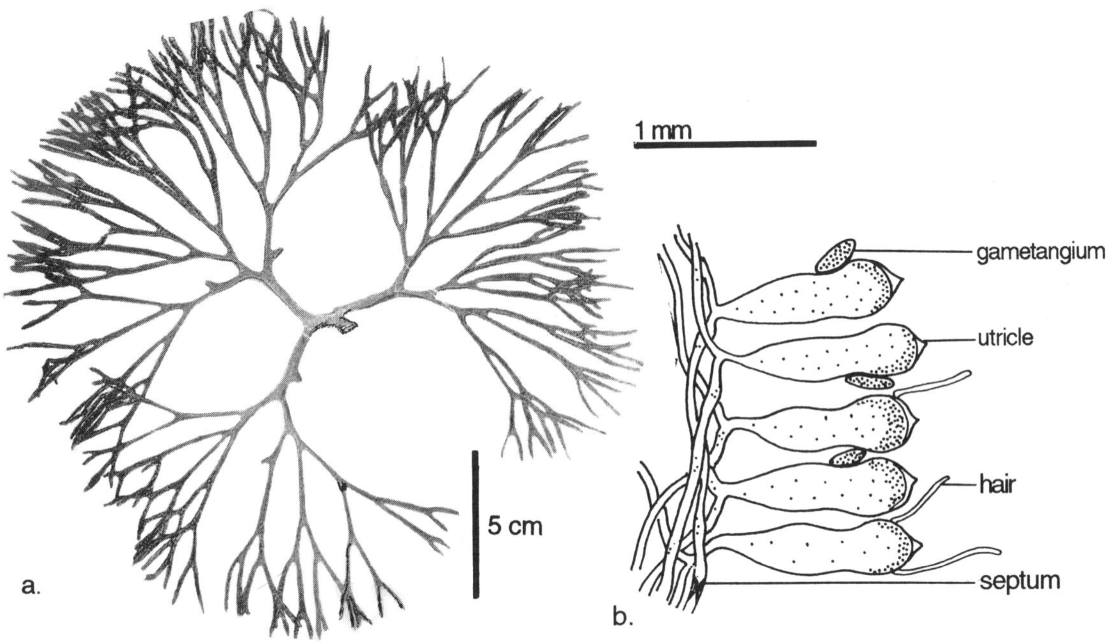

Fig 21. ***Codium fragile*** **ssp.** ***tomentosoides***
a. whole thallus (mature form); b. detail of utricles

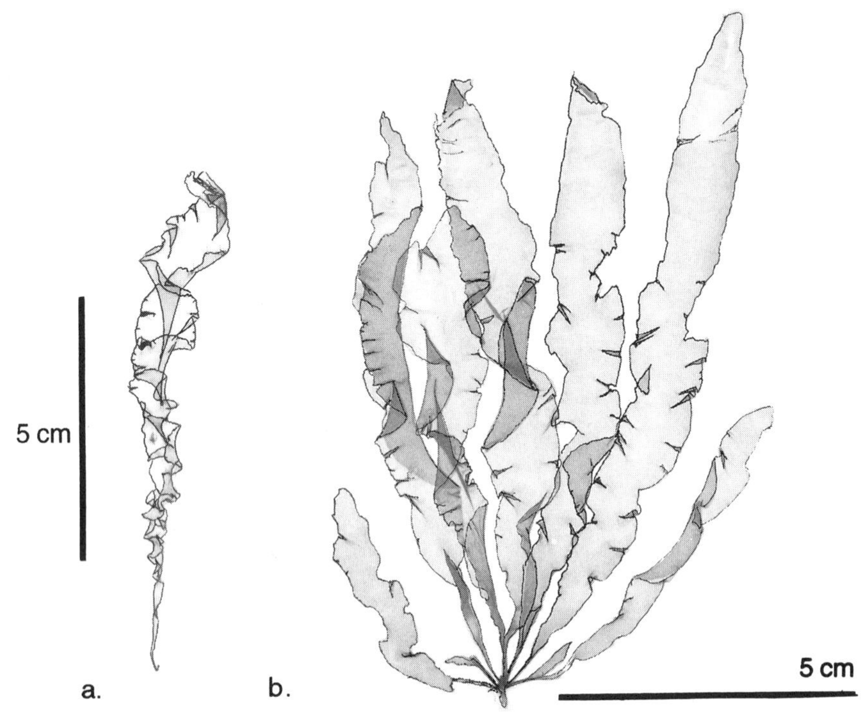

Fig. 22. ***Enteromorpha linza***
a. individual specimen showing undulated edges;
b. tuft

27. **Thin tubes, up to 15 cm long, tapering at base; dark green; cells in groups of 2 or 4, at least near base** (drawing below); thread-like below, no more than 1 mm wide above; growing in tufts or patches; found in late winter to early spring, on the open coast, in the low intertidal zone, attached to rocks and wood .... *Capsosiphon groenlandicum* (Fig. 23)

27. **Entangled, often twisted, tubular filaments of uniform diameter**; soft, up to 30 cm long, never branched; brownish yellow, to pale green; rounded cells (drawing below) in distinct longitudinal and transversal rows, often in groups of 2 or 4; found from spring to fall, in estuaries and around fresh water outlets, in the intertidal zone, attached to rocks, stones and mud ....*Capsosiphon fulvescens* (Fig. 24)

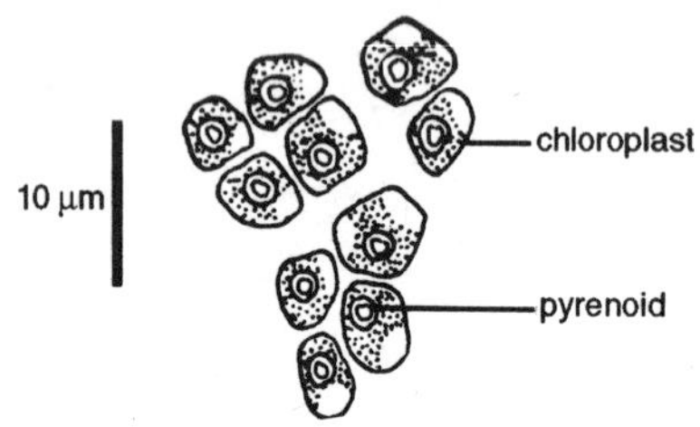

C. groenlandicum,
surface view of cells in midthallus

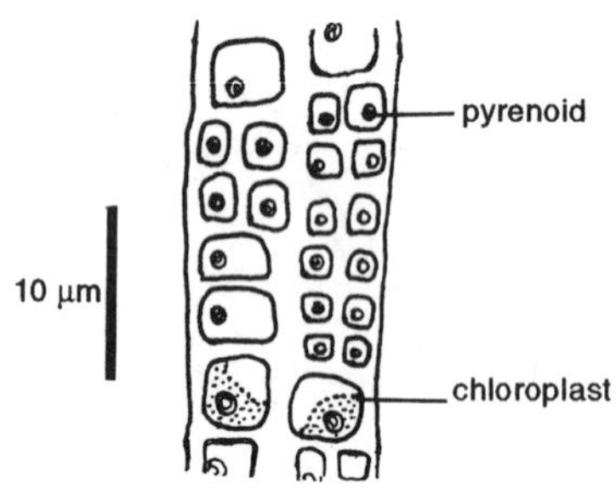

C. fulvescens,
surface view of cells in midthallus

28. **Tubes short, less than 5 cm long, attached by a small disc; cells small, less than 7 µm in diameter; chloroplast star shaped;** 1 pyrenoid per cell; thallus frequently contorted; yellow green; common year round, on the open coast and in estuaries; growing in tufts or carpet-like patches on rocks, wood or epiphytic on other algae, in the intertidal zone and just below .... *Blidingia minima* (Fig. 25)

28. **Tubes 5 cm long or more, and up to 5 cm in diameter; attached by rhizoids; cells 9-12 µm in diameter; chloroplast cup or hood-shaped and positioned on upper end of cell;** 1 pyrenoid per cell; thallus often inflated and contorted; very common year round, on the open coast and in estuaries, attached to any substrata, including other algae, in the intertidal zone and just below; also found in tide pools ....*Enteromorpha intestinalis* (Fig. 26)

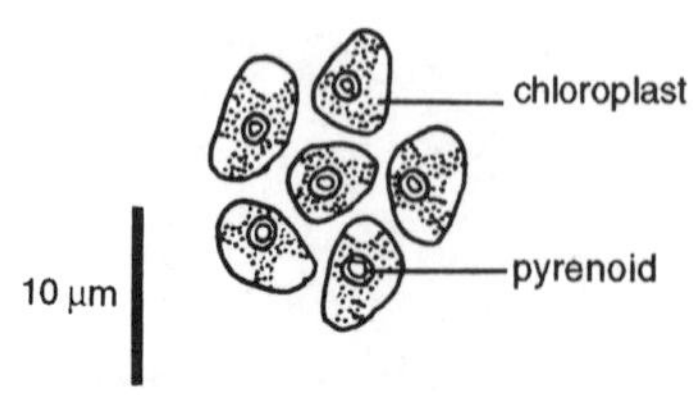

Blidingia minima,
surface view of cells in midthallus

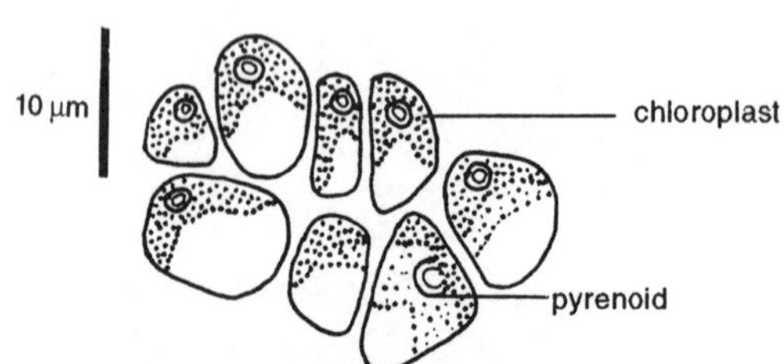

Enteromorpha intestinalis,
surface view of cells from the midthallus area,
with typical orientation of hood-shaped chloroplast

Fig. 23. ***Capsosiphon groenlandicum***
a. individual specimens; b. clump of filaments

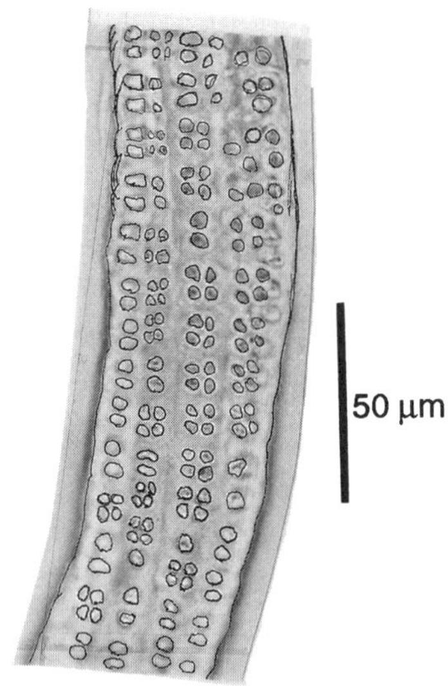

Fig. 24. ***Capsosiphon fulvescens***
detail of filament showing
cells in groups of 4s

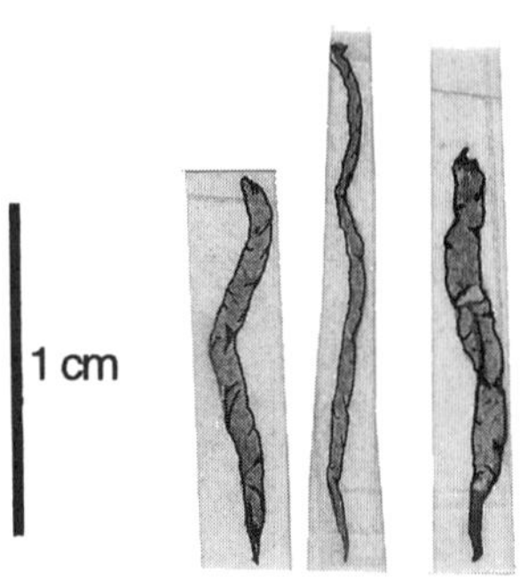

Fig. 25. ***Blidingia minima***
individual specimens

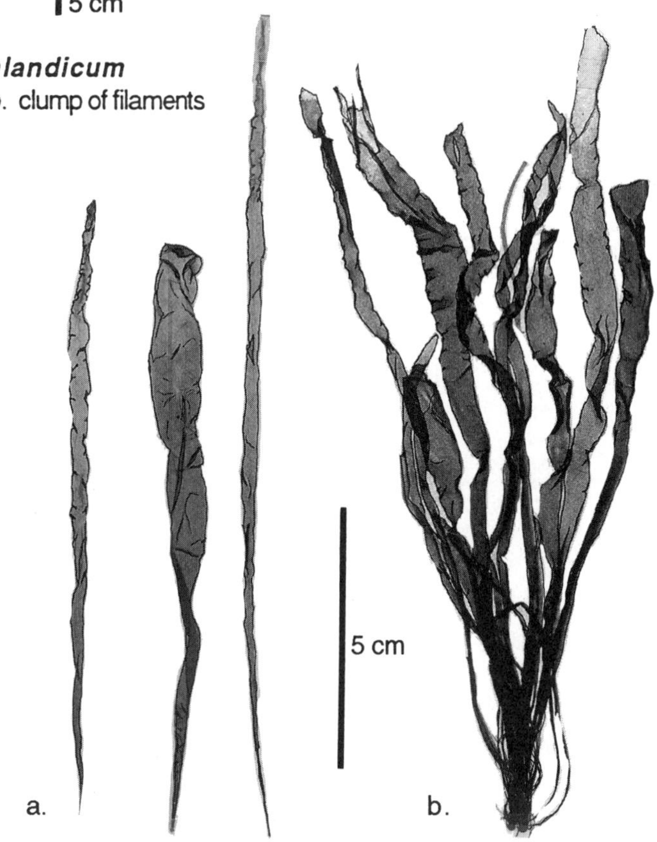

Fig. 26. ***Enteromorpha intestinalis***
a. individual specimens; b. tuft with whitish
reproductive patches at tips of tubes

29. **Cells polygonal or rounded, small, 9-12 μm in diameter, sometimes in rows; with mostly 1 pyrenoid per cell; branches often restricted to the base of the thallus**; axis somewhat compressed, with a long tapering base; up to 30 cm long; bright to dark green; branches contracted at base, similar to main axis; found year round, on the open coast, but more common in estuaries, in the intertidal zone and just below, attached to rocks and wood, or epiphytic on other algae .... *Enteromorpha compressa* * (Fig. 27)

29. **Cells somewhat square or rectangular;** shape of thallus extremely variable; cellular details needed for further identification 30

30. **Cells small (5-10 μm by 9-15 μm), in distinct rows, with mostly 1 pyrenoid per cell; main axis obvious, with few branches;** up to 50 cm long, by up to 1.5 cm wide; growing singly or in tufts; found year round, in the intertidal zone, attached to rocks and wood, or epiphytic on other algae; more common in estuaries .... *Enteromorpha prolifera* (Fig. 28)

30. **Cells 10 by 15 μm or larger, in longitudinal and transverse rows; more than 1 pyrenoid in most cells; chloroplast with irregular lobes** 31

* May be a subspecies of *Enteromorpha intestinalis* (Silva and Burrows, 1973; Burrows, 1991).

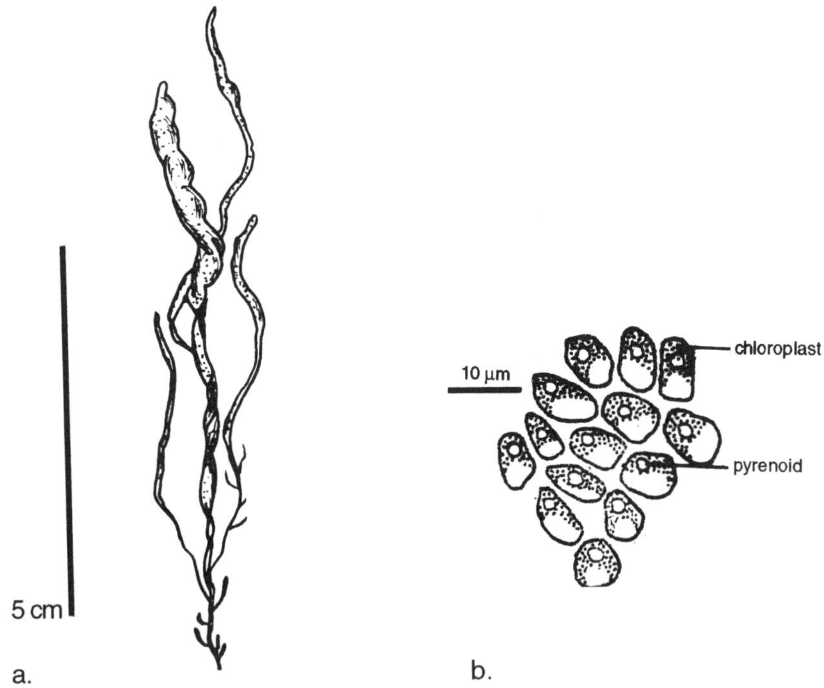

Fig. 27. ***Enteromorpha compressa***
a. single specimen; b. surface view of cells in midthallus

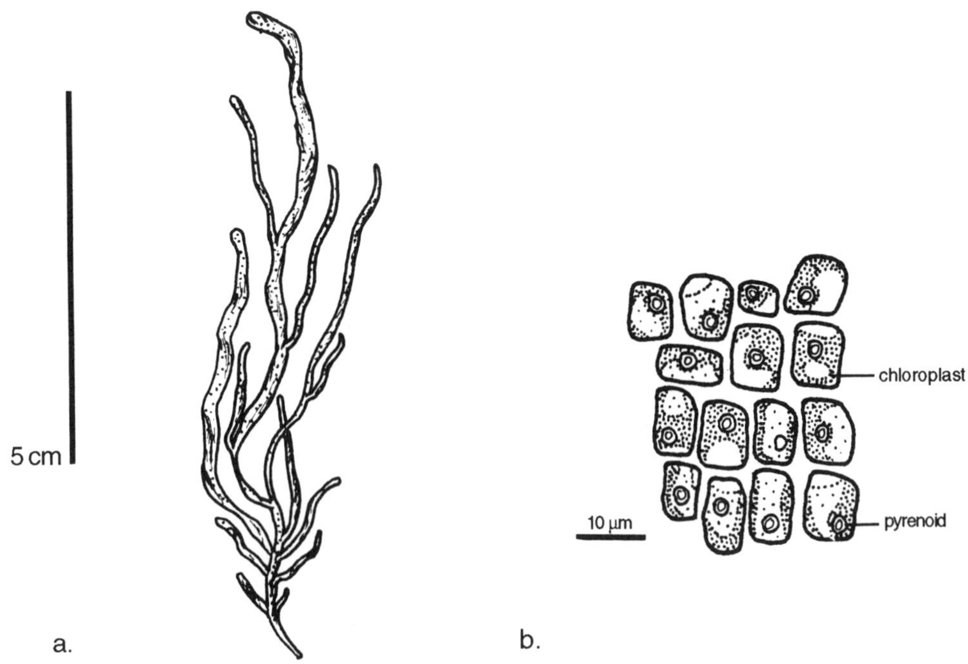

Fig. 28. ***Enteromorpha prolifera***
a. single specimen; b. surface view of cells in midthallus

30 (Green algae)

31. **Thallus sparsely branched or simple; cells 10-15 μm by 15-20 μm, 30 μm near base;** up to 300 mm long by up to 4 mm wide; cells in rows; chloroplast cylindrical; 1-2 (3) pyrenoid per cell in middle and at tip of thallus, 2-5 near base; found mostly in estuaries, in the summer, attached to rocks or wood, in protected areas, often mixed with other algae; more common north of Cape Cod .... *Enteromorpha flexuosa* ssp. *flexuosa* (Fig. 29)

31. **Branches abundant, ending in very fine, uniseriate tips**........................ 32

32. **Pyrenoids (1) 2-5 per cell; axes and branches very thin, hair-like;** cells 10-25 μm by 20-35 (50) μm, in longitudinal and transversal rows; chloroplast small, pale green, with lobes; branches often similar to the central axis; abundantly branched, with very fine, long uniseriate tips; found year round, on the open coast and in estuaries, in the mid to low intertidal zone, attached to rocks and shells, or epiphytic on other algae; often floating in brackish waters .... *Enteromorpha flexuosa* ssp. *paradoxa* (Fig. 30)

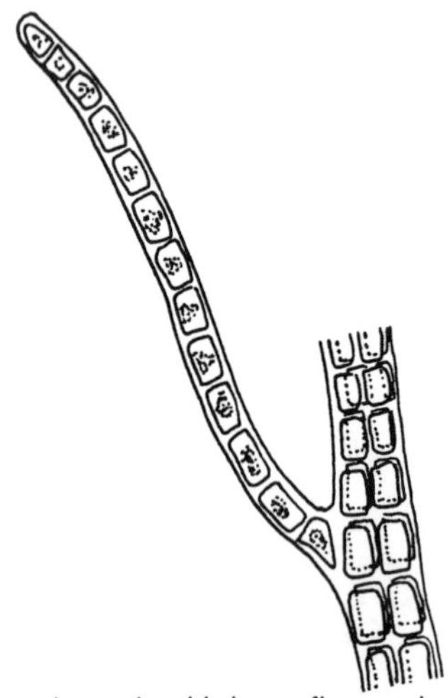

branch with long, fine, uniseriate tip

32. **Pyrenoids 2-6 per cell in middle and at tips of thallus, up to 10 (12) near base;** cells 10-35 μm by 15-40 (50) μm, in longitudinal and transversal rows usually visible in younger parts; chloroplast thin, delicate, plate-like; thallus up to 40 cm long by up to 2.5 cm wide, abundantly and repeatedly branched; branches fine, rebranched with uniseriate tips, 5 cells long or more; found mostly from spring to fall, attached to rocks, wood and shells, or epiphytic on other algae; often forming large, free floating mats, in quiet estuaries, particularly when older .... *Enteromorpha clathrata* (Fig. 31)

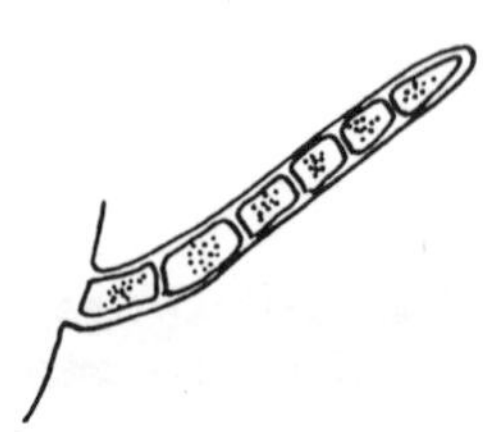

branch with fine, uniseriate tip

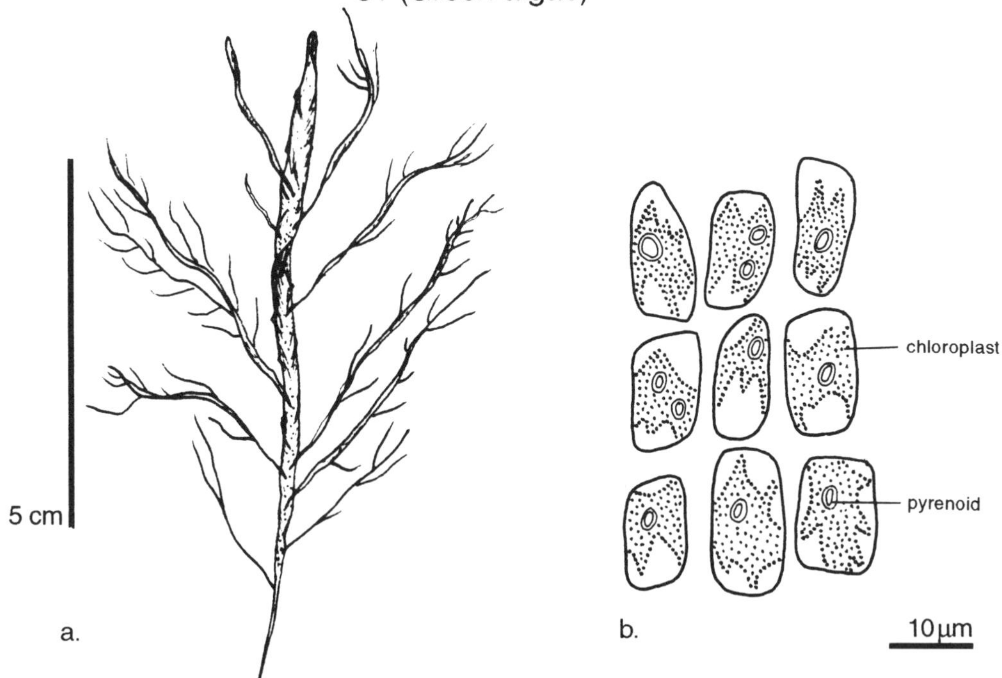

Fig. 29. ***Enteromorpha flexuosa*** **ssp.** ***flexuosa***
a. single specimen; b. surface view of cells in midthallus

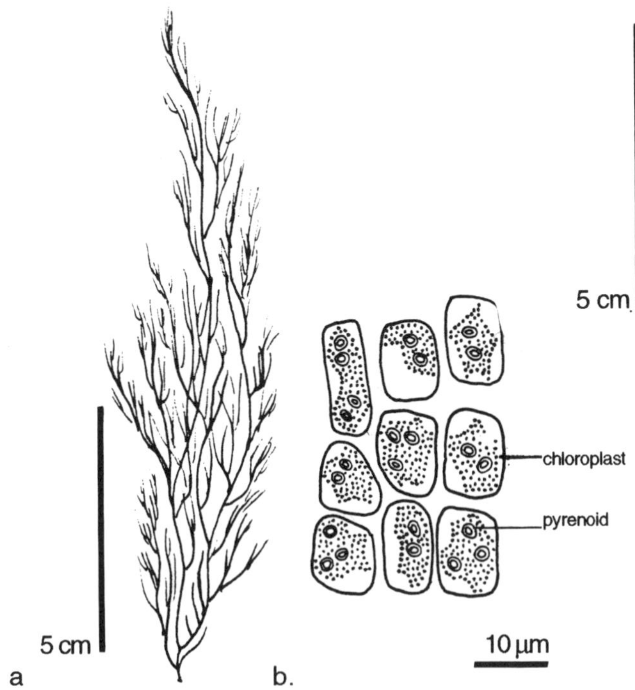

Fig. 30. ***Enteromorpha flexuosa*** **ssp.** ***paradoxa***
a. single specimen
b. surface view of cells in midthallus

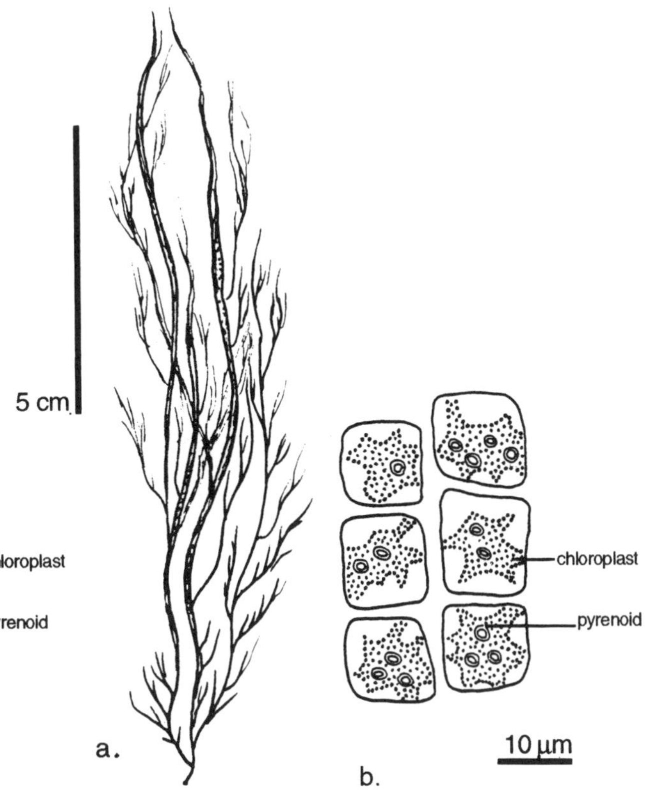

Fig. 31. ***Enteromorpha clathrata***
a. single specimen
b. surface view of cells in midthallus

32 (Green algae)

33. **Blade small, less than 1 cm high**; often curled, dark green; cells spaced in 4s or multiples of 4; common year round, mostly on the open coast, forming green patches at, or above the high tide line, on rocks, particularly in shaded areas between boulders and in crevices; often growing around areas of bird droppings .... *Prasiola stipitata* (Fig. 32)

33. **Blade otherwise** 34

34. **Blade 2 cell layers thick in cross-section (drawing below); firm, bright green; up to 50 cm long** (1 m or more in quiet, eutrophic waters); blade rounded, oval or elongated, often undulated at edges; sometimes lobed or jagged at tip; may be perforated with holes of various sizes; growing singly, in tufts or patches; very common year round, on the open coast and in estuaries, in the intertidal and subtidal zones, down to -24 m or more, attached to rocks, shells, wood or epiphytic on other algae; often drifting in large masses in protected, eutrophic areas. One of the most common seaweeds in New England .... *Ulva lactuca* (Fig. 33)

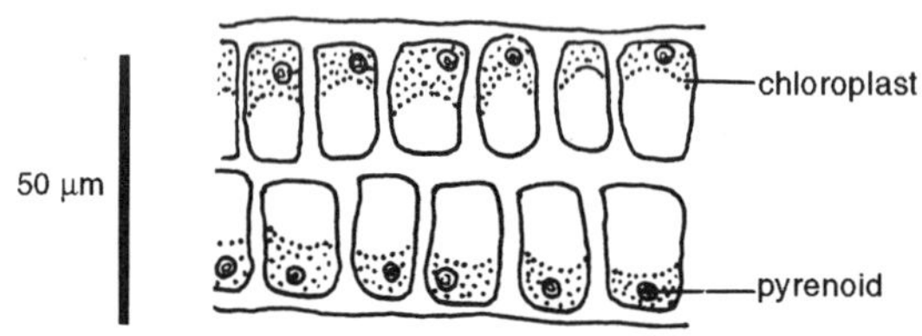

Ulva lactuca, cross-section

34. **Blade 1 cell layer only in cross section;** firm or delicate................. 35

35. **Blade dark green, flat and firm; turning dark brown when dried;** up to 40 cm long; a single layer of cells in cross-section (drawing below); 65-75 µm thick; gradually expanding from base, sometimes lobed; common year round, from Cape Cod northward, on the open coast and in estuaries, in the lower intertidal zone and below, on rocks, shells and other algae and in tide pools; often found in large floating masses ..*Ulvaria obscura* var. *blyttii* * (Fig. 34)

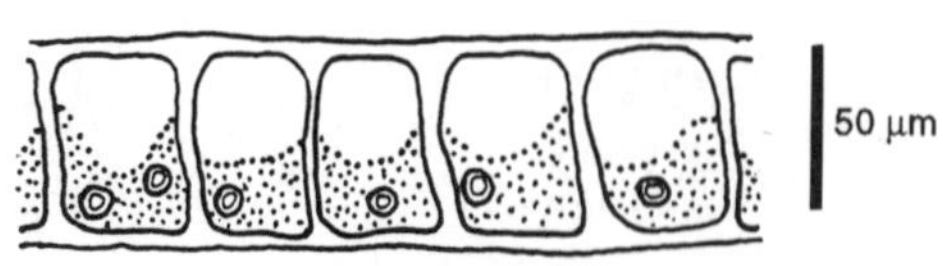

U. obscura var. *blyttii* , cross-section

35. **Blade thin and delicate, light green**; up to 30 cm long; 1 cell layer thick in cross-section; found from late winter to early summer, occasionally in fall 36

* Nomenclature uncertain; as *Monostroma obscurum* in Burrows (1991).

33 (Green algae)

Fig. 33. ***Ulva lactuca***
a. lobed blade; b. ruffled blade with perforations

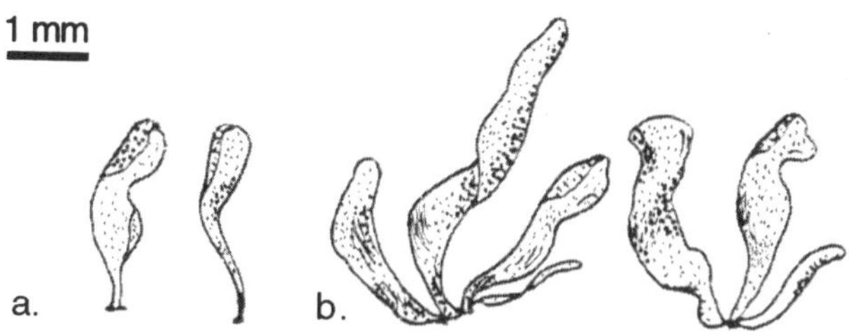

Fig. 32. ***Prasiola stipitata***
a. individual specimens; b. clumps

Fig. 34. ***Ulvaria obscura* var. *blyttii***
whole specimen showing dark color and ruffled edges

36. **Blade elongated, with deeply undulated edges; very delicate and fragile**; up to 5 cm wide by 30 cm long; 20-32 µm thick; common from winter to late spring, on the open coast and in estuaries, attached to rocks or epiphytic on other algae, around the low tide level ... . *Protomonostroma undulatum* * (Fig. 35)

36. **Blade otherwise** 37

37. **Cell walls thick (diagram below); cells rounded or with rounded corners, sometimes grouped in 2s or 4s**; growing in thick tufts; sac-like when young, later splitting to wide, somewhat ruffled segments, up to 10 cm long; may be larger (up to 50 cm) when floating; blade 20-40 (60) µm in cross-section; common in spring and summer (but also recorded in fall); in estuaries, salt marshes and quiet, shallow, brackish waters, on many substrata, common on eelgrass south of Cape Cod; extends to the upper intertidal north of Cape Cod .... *Monostroma oxyspermum* (Fig. 36)

37. **Cell walls thin (diagram below)** 38

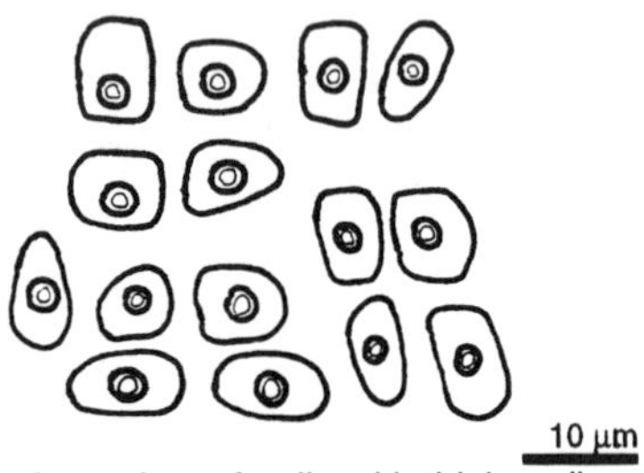

surface view of cells with thick walls from midthallus area

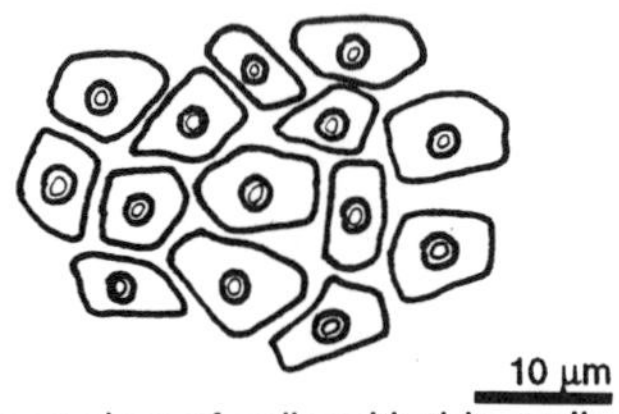

surface view of cells with thin walls from midthallus area

38. **Blade attached, sac-like when young, opening into a flat sheet at maturity, splitting into segments down to the base; cells 11-17 (22) µm, one pyrenoid per cell** (drawing below); thallus 18 (35) µm in cross-section, up to 15 cm tall, light olive-green; common from fall to spring, on the open coast and in estuaries, in the low intertidal, in tufts or large patches, attached to rocks, stones and shells, or epiphytic on other algae; also found in tide pools .... *Monostroma grevillei* (Fig. 37)

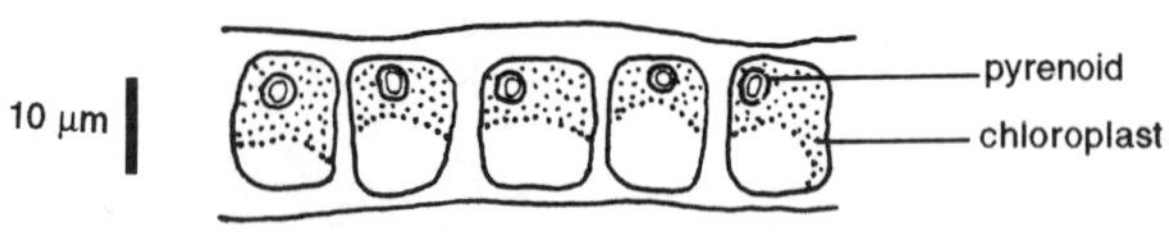

Monostroma grevillei, cross-section

Kornmannia leptoderma, cross-section

38. **Blade often floating; cells small, 5-10 µm in diameter, no visible pyrenoid**; thallus 10 µm or less in cross-section (drawing above), up to 10 cm long; found in late winter and early spring (extends into September, north of Cape Cod); growing mostly in estuaries, around the low tide level and just below, attached to rocks, or epiphytic on other algae and eelgrass; also found in tide pools .... *Kornmannia leptoderma* * * (Fig. 38)

* Nomenclature uncertain; as *Monostroma undulatum* in Burrows (1991).

* * Nomenclature uncertain; possibly included in *Monostroma* (Burrows,1991).

35 (Green algae)

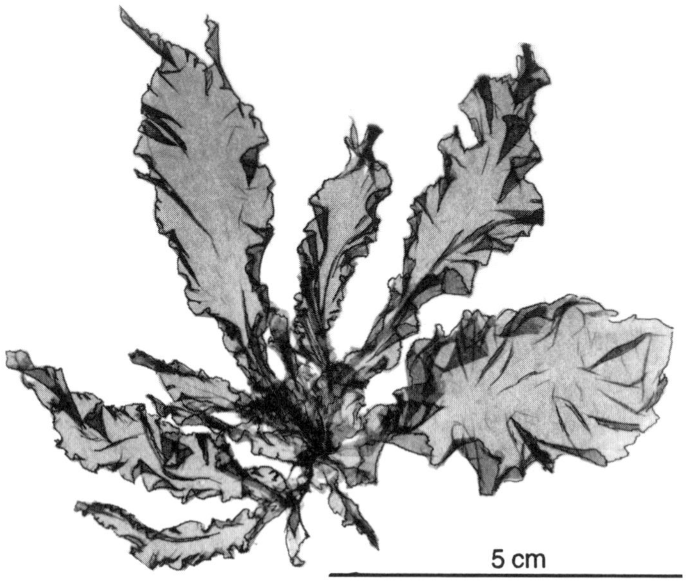

Fig. 35. ***Protomonostroma undulatum***
tuft showing blades with undulated edges

Fig. 36. ***Monostroma oxyspermum***
individual specimens showing blades with ruffled segments

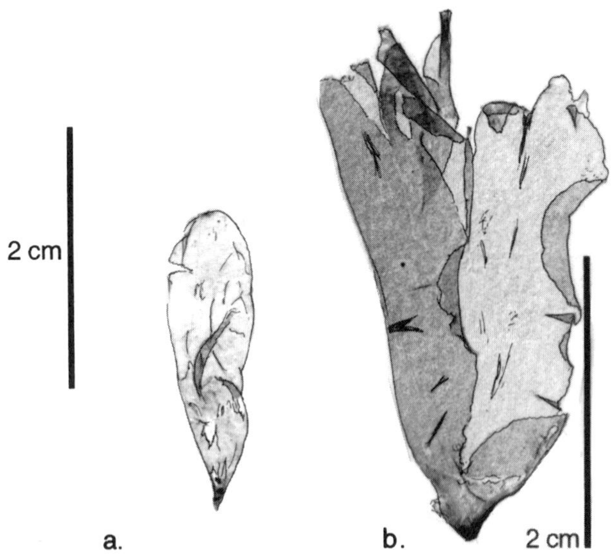

Fig. 37. ***Monostroma grevillei***
a. young, sac-like specimen
b. mature specimen

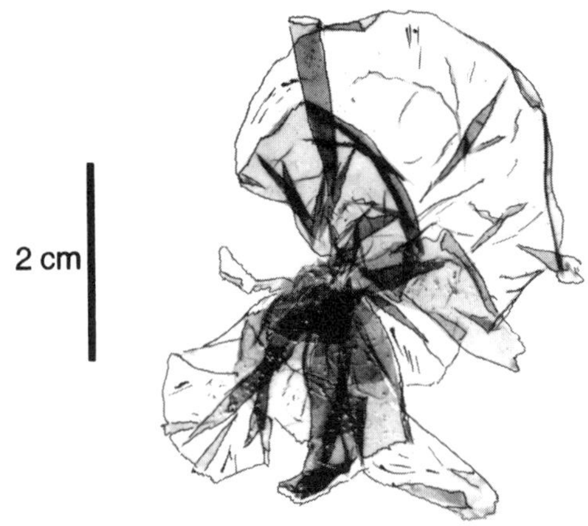

Fig. 38. ***Kornmannia leptoderma***
individual specimen

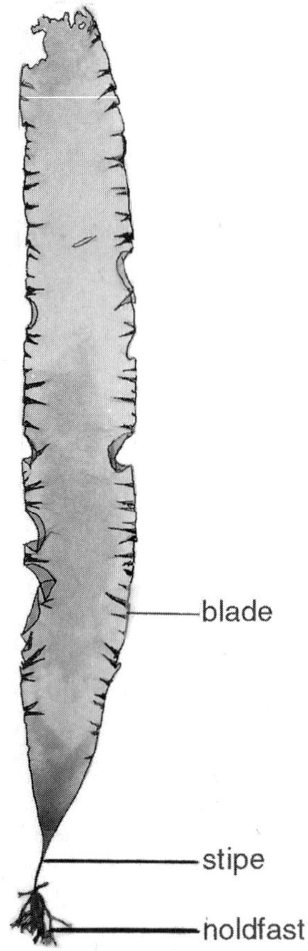

Laminaria saccharina

KEY TO THE PHAEOPHYTA

Major diagnostic characters

	Page	Step
Crusts, spots, or felt-like mats	38-40	2-10
Short tufts of uniform size	42-46	12-18
Uniseriate filaments	46-50	20-26
Globular masses	50	27
Unbranched tubes or cords	52	29-32
Unbranched blades (may be divided longitudinally)	54-58	34-42
Flattened branches (3-10 mm wide)	60-62	44-49
Worm-like (soft and gelatinous)	64	50-51
With whorls of branchlets (as on a pipe cleaner)	64	53
Covered with hairs	66	54-56
Bushy, branched, without conspicuous hairs	66-68	57-62

1. **Brown spot, patch, crust, or felt-like mat; on rocks, shells or other algae**; consisting of upright filaments that arise from a single layer of basal horizontal filaments attached to the substrate; individual thallus generally no more than 3 mm thick; in some species, older specimens with overlapping lobes, may reach 2 cm in thickness; microscopic examination of vertical sections may be needed for identification 2
1. **Thallus upright, or floating when in water;** longer than 3 mm 11

2. **Felt-like spots, tufts or patches;** attached to other algae or to shells 3
2. **Crust or spot not felt-like;** attached to rocks, shells or other algae 4

3. **Endophytic, felt-like spots or patches; basal filaments extending between host cells; plurilocular sporangia uniseriate** (see drawing); upright filaments unbranched or with few branches, generally 1 mm long or less, but reaching up to 1 cm; uncommon; found mostly on *Laminaria* spp. and, since july 1996, on *Grateloupia doryphora*; from Rhode Island northward, from february to december, on the open coast, in the subtidal zone .... *Laminariocolax tomentosoides* (Fig. 1)
3. **Epiphytic, felt-like tufts or patches, found on other algae and occasionally on shells and rocks; basal filaments limited to substrate surface; plurilocular sporangia pluriseriate;** upright filaments branched, up to 1 (3) mm long; hairs with basal sheaths present; found occasionally only, from Cape Cod northward, from spring to fall, in the intertidal zone of the open coast ... *Protectocarpus speciosus* (Fig. 2)

4. **Small brown discs, .5-1 mm wide, generally epiphytic on other algae; individual discs occasionally merging; upright filaments loosely packed;** cushion-like, inconspicuous 5
4. **Crusts of various sizes, mostly attached to rocks; upright filaments tightly packed** 7

5. **Solitary, light brown discs, epiphytic on *Dumontia contorta* only; basal filaments spreading irregularly and producing rhizoids into host;** discs up to 1 mm in diameter; with many upright filaments, hairs and sporangia; common from Cape Cod northward, in low intertidal tide pools, from late winter to fall ...... *Ulonema rhizophorum* (Fig. 3)
5. **Brown discs epiphytic on various algae; rhizoids absent** .... (*Myrionema* spp. ♦2 of 4) 6

6. **Brown discs, 2-3 mm in diameter; epiphytic on *Laminaria* spp. only; upright filaments, 100-140 µm long, linear (with the same diameter throughout);** found from late winter to fall, on the open coast, in low intertidal tide pools and just below the low tide level .... *Myrionema corunnae* (Fig. 4)
6. **Brown discs, up to 1 mm in diameter; attached to various algae; upright filaments, 50-100 µm long, club-shaped;** common during the summer, from Cape Cod southward, but only occasionally to the north; growing in low intertidal tide pools and in the shallow subtidal zone, from mid-winter to fall; also attached to oyster shell .... *Myrionema strangulans* (Fig. 5)

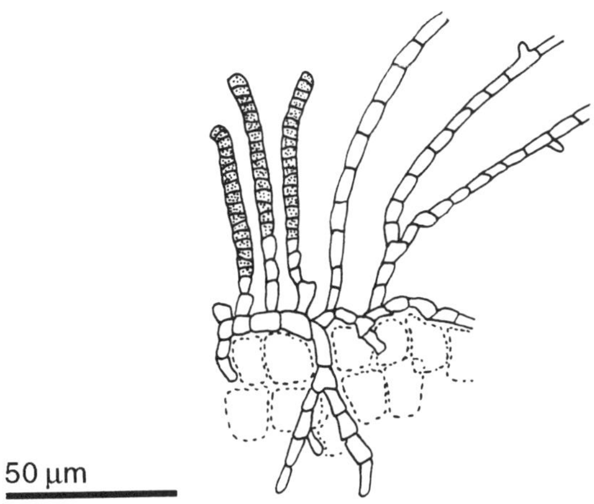

Fig. 1. ***Laminariocolax tomentosoides***
portion of colony with uniseriate sporangia and basal filaments growing into host

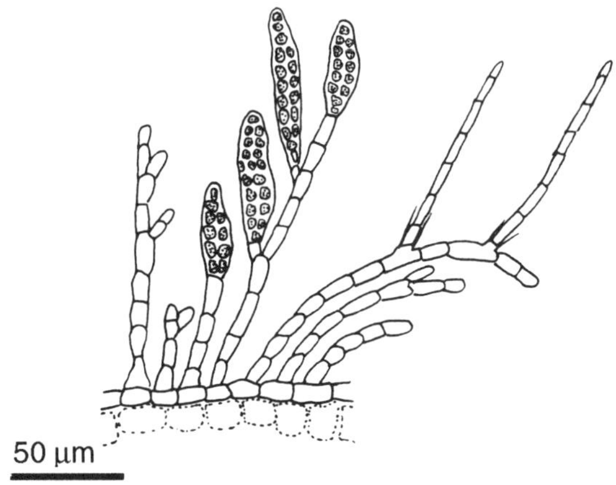

Fig. 2. ***Protectocarpus speciosus***
portion of colony with pluriseriate sporangia and basal filaments on surface of host cells

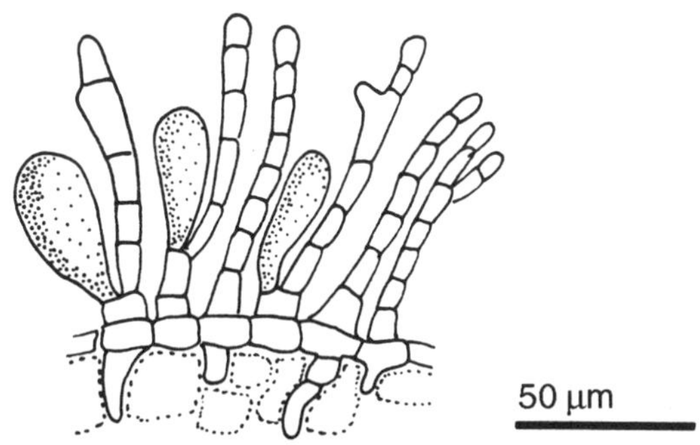

Fig. 3. ***Ulonema rhizophorum***
portion of colony with unilocular sporangia and rhizoids growing into host

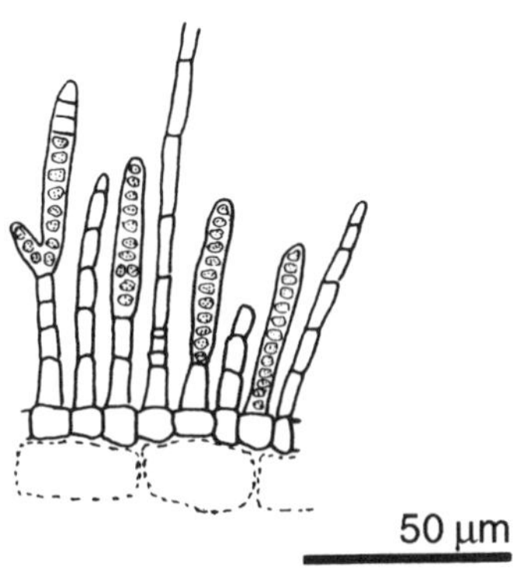

Fig. 4. ***Myrionema corunnae***
portion of colony with plurilocular sporangia and linear upright filaments

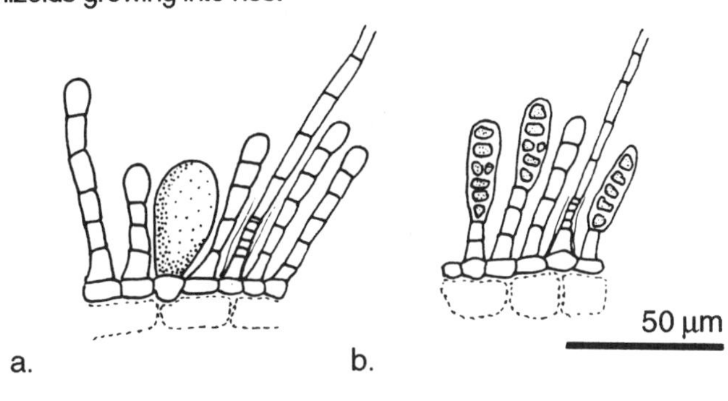

Fig. 5. ***Myrionema strangulans***
a. portion of colony with unilocular sporangium and somewhat club-shaped upright filaments;
b. with plurilocular sporangia

(Fig. 1, redrawn after Cardinal, 1964; Figs. 2, 3, 4 and 5a and b, redrawn after Fletcher, 1987)

7. **Cells with 4-7 irregular, disc-shaped plastids** (drawing below); large, rounded brown crust, 1 mm thick by 8 cm or more in diameter; upright filaments tightly packed; firmly attached to rocks or shells, in the subtidal zone, down to -15 m; common year round, from Cape Cod northward, on the open coast and in estuaries; (♦1 of 3). A similar species, *P. paradoxum* (soft crust with filaments separating easily and cells with 7-9 plastids) can also occasionally be found on rocks, in the subtidal zone, in southern Massachusetts (Sears and Wilce, 1973) .... *Pseudolithoderma extensum* (Fig. 6)

7. **Cells with a single plate-like plastid,** (drawing below) 8

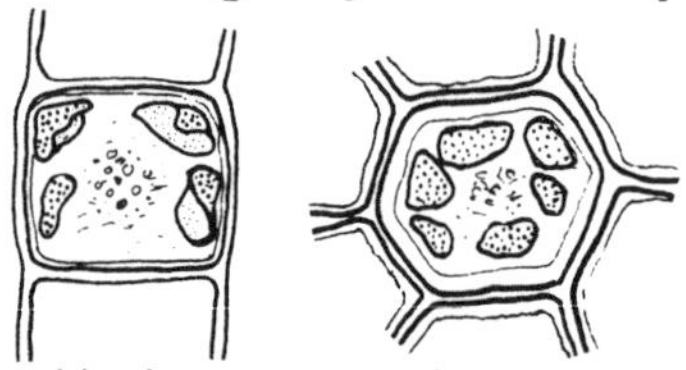

side view surface view

irregular disc-shaped plastids

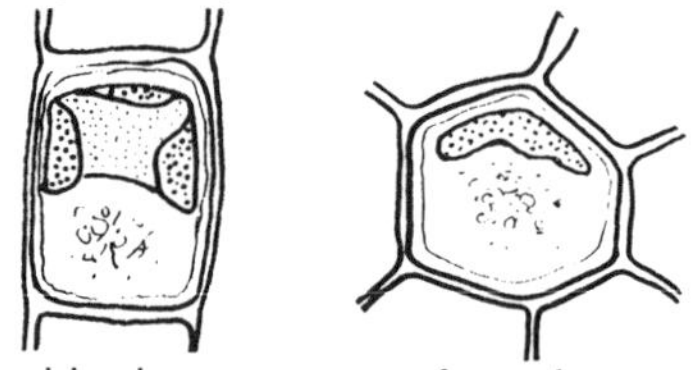

side view surface view

single plate-like plastid

8. **Crusts soft, occasionally spongy, firmly attached and difficult to remove from substrate; filaments spreading easily with pressure between slide and cover slip** 9

8. **Crust rigid, rough and brittle, particularly when old, easily scraped off subtrate; filaments do not spread easily with pressure; young discs merge into large patches** (*Ralfsia spp.* ♦2 of 3) 10

9. **Upright filaments tightly clustered, with frequent branching;** uncommon; found year round, on the open coast and in estuaries, from around the low tide level, down to -26 m, attached to rocks. This crust is a stage in the life cycle of *Petalonia sp.* and *Scytosiphon sp.* and is no longer considered a valid taxon .... *"Stragularia clavata"* (Fig. 7)

9. **Upright filaments loosely clustered; rarely branched;** common year round, attached to rocks, in the subtidal zone; more common on the open coast than in estuaries; southern Massachusetts is its southern limit .... *Petroderma maculiforme* (Fig. 8)

10. **Crusts, dark brown to black; 2-6 cm in diameter; with rounded overlapping lobes, concentrically and radially lined, resembling a bracket fungus; when seen in a vertical section, the filaments are distinctly curved in an upward layer and in a downward one;** older specimens with many overlapping lobes, may reach 2 cm in thickness; rare south of Cape Cod; more common to the north, in the intertidal and subtidal zones, down to -8 m; found year round, on the open coast, attached to rocks; .... *Ralfsia fungiformis* (Fig. 9)

10. **Crusts, dark brown to mottled orange; .5-10 cm in diameter; 1-2 mm thick; upright filaments curved upward only, when seen in vertical section;** circular and smooth when young, becoming irregular, rough and brittle with age; common year round, throughout the intertidal zone and below, down to -20 m, forming large patches, mostly on rocks, on the open coast and in estuaries; this is the most common of the brown crusts found throughout New England .... *Ralfsia verrucosa* (Fig. 10)

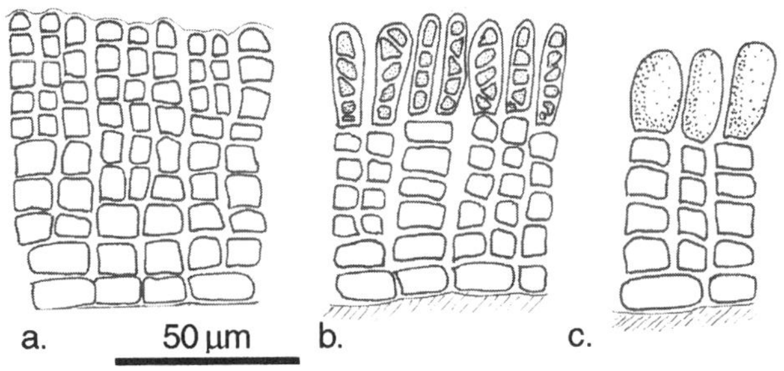

Fig. 6. ***Pseudolithoderma extensum***
a. portion of crust with tightly packed vertical filaments; b. with plurilocular sporangia; c. with unilocular sporangia

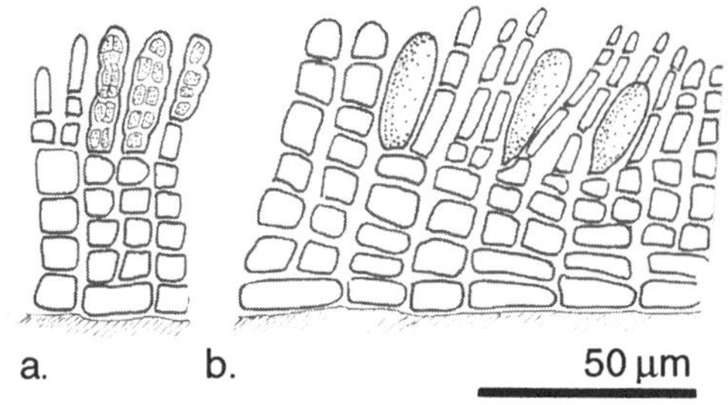

Fig. 7. ***"Stragularia clavata"***
a. portion of crust with plurilocular sporangia; b. with unilocular sporangia

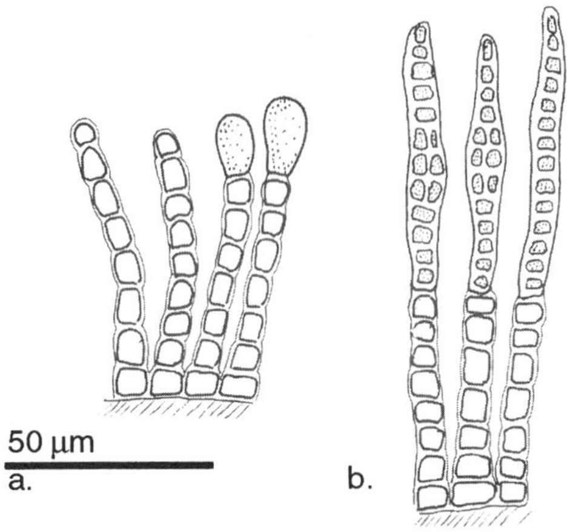

Fig. 8. ***Petroderma maculiforme***
a. portion of crust with loosely clustered filaments and unilocular sporangia
b. with plurilocular sporangia

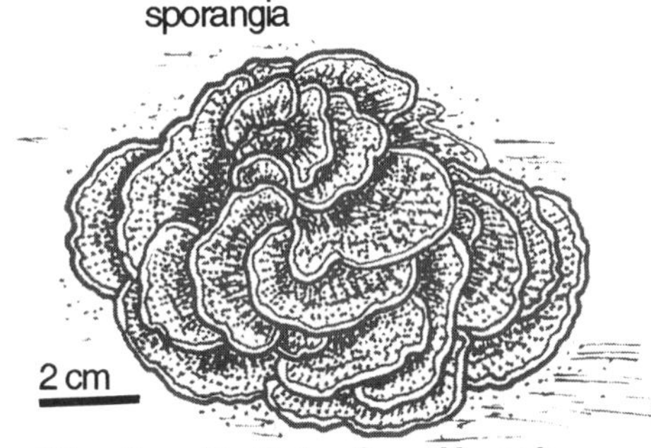

Fig. 9. ***Ralfsia fungiformis***
surface view of crust showing overlapping lobes

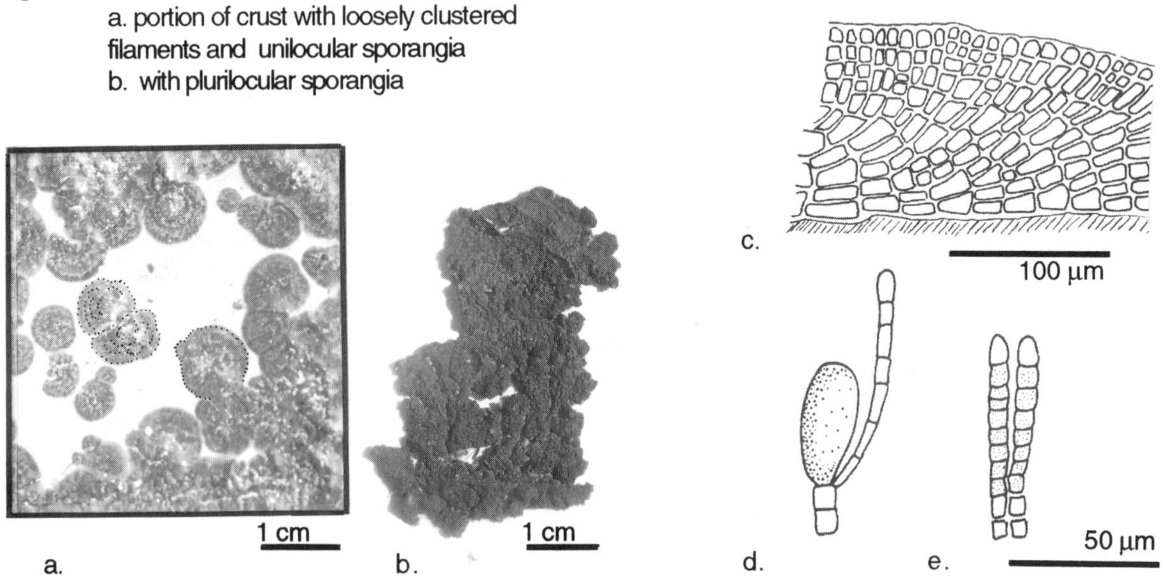

Fig. 10. ***Ralfsia verrucosa;*** a. young colonies on rocks; b. older colony detached; c. portion of colony with curved filaments; d. unilocular sporangium; e. plurilocular sporangium

(Figs. 6a, b and c; 7a and b; 8a and b; 10c, d and e; redrawn after Fletcher, 1987)

11. **Short tufts of relatively uniform height, attached to other algae**; soft or stiff; solitary or sometimes crowded and occasionally forming a sheath around the host 12
11. **Thallus otherwise** 19

12. **Tufts somewhat stiff, up to 5 (8) cm high; axes and branches uniseriate when young, becoming pluriseriate (more than one cell wide; diagram below), with cells in regular transverse rows, at least in upper parts;** with or without distinct axes; with a conspicuous, often dark, large growth cell, at tips of branches; (*Sphacelaria* spp., ♦6 of 9) 13
12. **Tufts or clumps soft, no more than 2.5 cm tall; uniseriate filaments branched only near base;** no conspicuous apical cell 18

Sphacelaria spp.
typical cell pattern

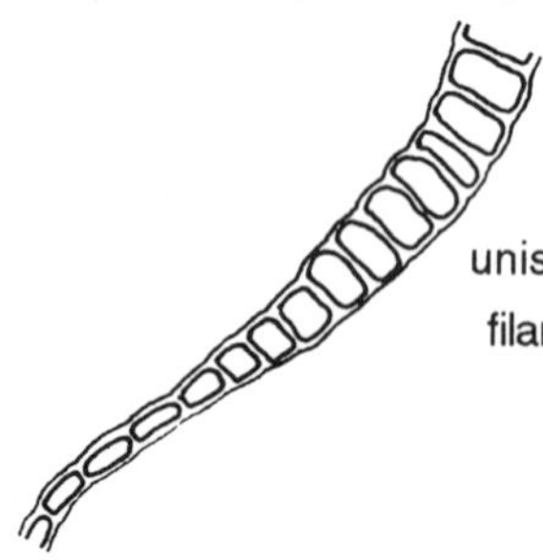

uniseriate
filament

13. **Propagules usually present (Fig. 12e);** tufts usually 1 to 2 cm high, but up to 5 cm, solitary or clustered; branching irregular and abundant; cells in main axes, in distinct transverse rows; surface cells not redivided transversally; terminal cells may be large and dark; common year round, in the intertidal and subtidal zones, down to -15 m, on rocks, under larger algae, or epiphytic mostly on fucoids, on the open coast and in estuaries; often free floating in tangled masses or in rounded tufts. This is the most common of the *Sphacelaria* species .... *Sphacelaria cirrosa* (Fig. 11)
13. **Propagules absent;** surface cells often redividing transversally 14

14. **Branching feather-like, at least in upper parts;** base of axes covered by a cortical layer of rhizoids; holdfast a thick basal disc; axes wiry, up to 10 cm long 15
14. **Branching mostly irregular** (alternate, opposite or clustered on the same branch); axes slender; if rhizoids are present, they are loosely arranged 16

15. **Main axes 250-700 µm near base, heavily corticated by rhizoids which grow from any surface cells;** tufts bushy and wiry, solitary or in clusters; feather-like branching often restricted to upper parts; uncommon; found from Cape Cod northward, on the open coast and in estuaries, attached to rocks, stones, shells and occasionally other algae, in low tide pools and in the subtidal zone, down to -20 m; also found free floating .... *Sphacelaria plumosa* (Fig. 12)
15. **Main axes 70-100 µm near base, only partially and irregularly corticated, with rhizoids developing only in the plane of the branches;** tufts shrubby but not wiry, solitary or in clusters; feather-like branching extending almost to the base; rare north of Cape Cod, more common to the south, mostly on the open coast, attached to rocks, sand and occasionally other algae, in the subtidal zone, down to -20 m .... *Sphacelaria plumigera* (Fig. 13)

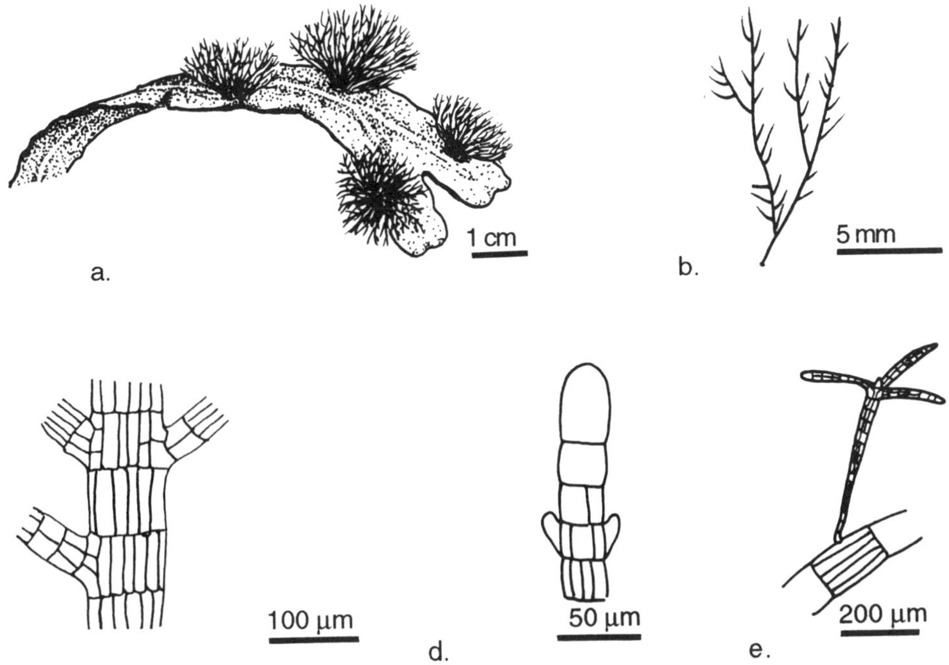

Fig. 11. ***Sphacelaria cirrosa;*** a. tufts attached to host; b. detail of branching pattern; c. cell pattern showing pluriseriate growth; d. terminal cell; e. propagule

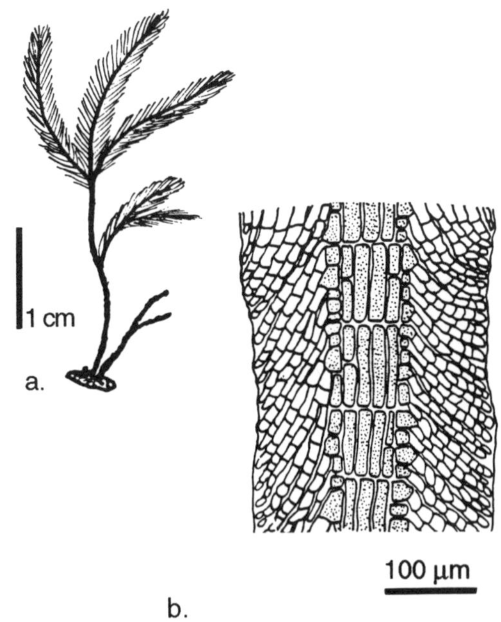

Fig. 12. ***Sphacelaria plumosa***

a. feather-like branching pattern
b. longitudinal section of axis heavily corticated by rhizoids

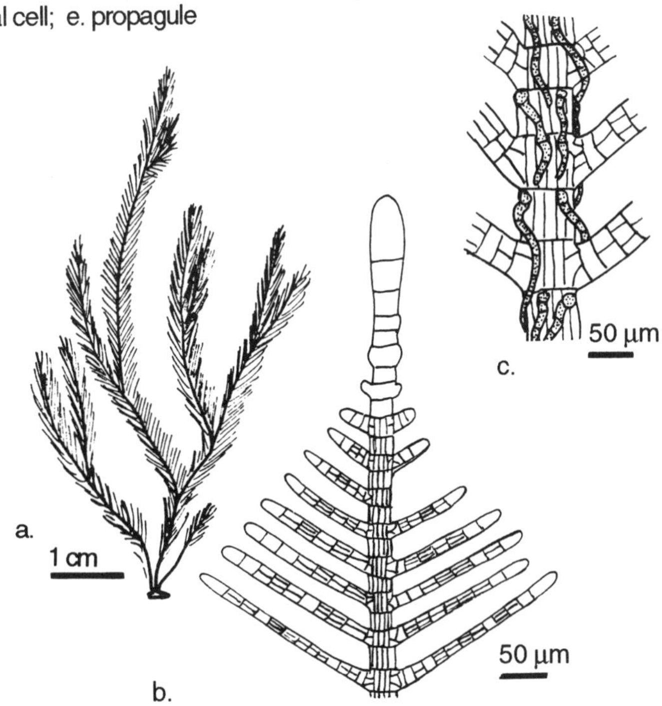

Fig. 13. ***Sphacelaria plumigera***

a. feather-like branching pattern
b. detail with opposite branching
c. longitudinal section of axis with rhizoids growing in the plane of the branches

(Figs. 12b, 13b and c, redrawn after Prud'homme van Reine, 1982)

16. **Pericysts present (Fig. 15b); tufts generally crowded and forming extensive carpet-like mats;** numerous main axes, up to 2 cm long, with scattered, irregular branches; base of axes occasionally with a few rhizoids; common year round, on the open coast and in estuaries, mostly in the intertidal zone, in somewhat sheltered areas, attached to rocks, sand and mud; often found in tide pools and in salt marshes .... *Sphacelaria radicans* (Fig. 14)

16. **Pericysts absent; tufts solitary or sometimes crowded, but not forming extensive, solid mats** 17

17. **Tufts shrubby and broom-like, or elongated and asymmetrical, up to 8 cm long; rhizoids often present near base, forming a loosely organized sheath;** tufts stiff, solitary or in clusters; found year round, from Cape Cod northward, in low tide pools, and in the subtidal zone, down to -30 m, on the open coast and in estuaries, attached to rocks, shells, wood or epiphytic on other algae; also commonly found free floating .... *Sphacelaria arctica* (Fig. 15)

17. **Tufts short, shrubby but not broom-like, up to 2 cm long; few rhizoids present, never forming a sheath;** tufts with broadly spreading main axes and few irregular branches; solitary or forming small mats; abundant in some locations; found throughout the intertidal zone and in the subtidal zone down to -18 m, attached to rocks; also growing in sandy tide pools .... *Sphacelaria racemosa* (Fig. 16)

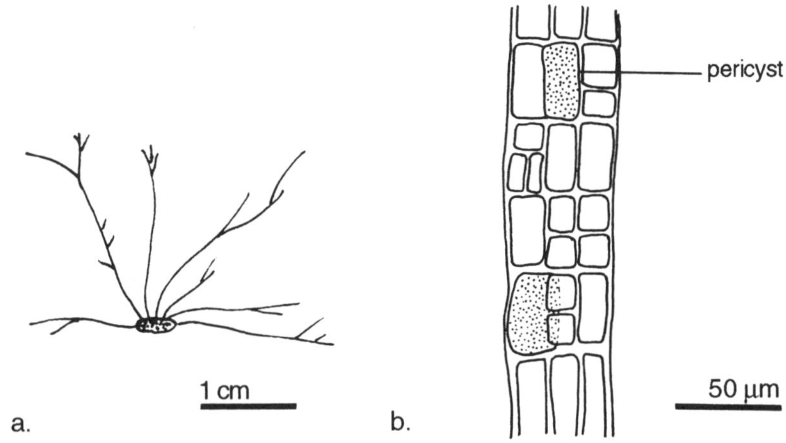

Fig. 14. ***Sphacelaria radicans***
a. branching pattern
b. detail of axis showing pericysts

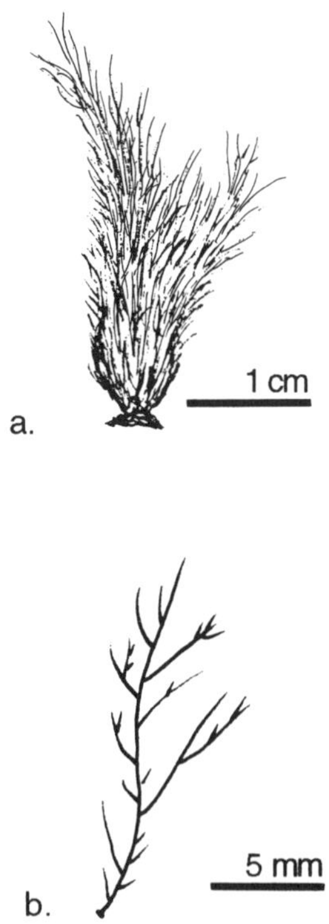

Fig. 15. ***Sphacelaria arctica***
a. broom-like tuft
b. branching pattern

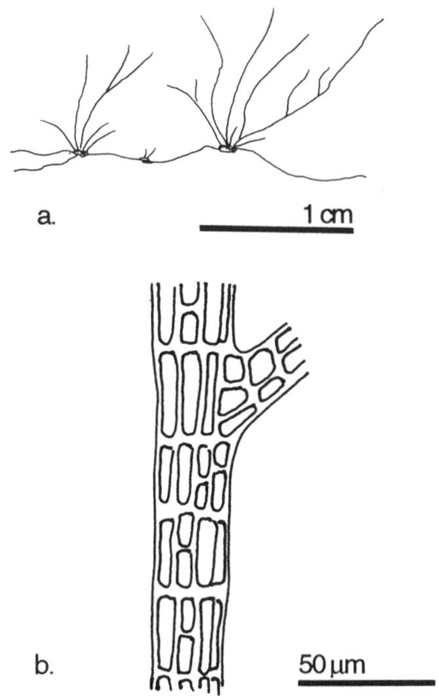

Fig. 16. ***Sphacelaria racemosa***
a. broadly spreading tuft with few axes
b. detail with branch

(Figs. 14a and b; 15a and b; 16a and b; redrawn after Prud'homme van Reine, 1982)

18. **Tuft growing from a small cushion-like knob; plurilocular structures terminal, on short branches at base of filaments;** erect tufts of uniseriate filaments, branched at base only; 1-1.5 (3) cm tall; common year round, on the open coast and in estuaries, in the intertidal zone and below, usually attached to *Fucus* spp., but also found on *Ascophyllum nodosum*; (♦1 of 3) .... *Elachista fucicola* (Fig. 17)

18. **Tuft growing from a small superficial holfast; plurilocular structures in patches or forming a sheath around the filaments;** erect tufts of uniseriate filaments, narrowing to base; up to 2.5 cm long; uncommon; found in winter and spring, epiphytic on eelgrass, *Ascophyllum nodosum* and other algae .... *Halothrix lumbricalis* (Fig. 18)

19. **Thallus very soft, fine and hair-like, forming irregular clumps or cloudy masses when in water; formed by highly branched filaments, usually uniseriate** (a single series of cells attached end to end, drawing below) **throughout the thallus, at least when young, with no visible main axis**; microscopic characteristics are needed to identify these species 20

19. **Thallus with 1 or more visible main axis; each axis, many cells wide** 27

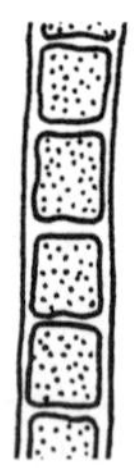

unbranched, uniseriate filament

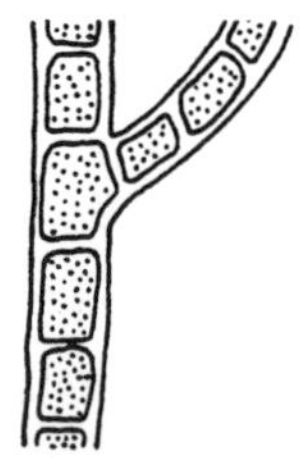

branched, uniseriate filament

20. **Plastids band-shaped** (drawing below); plurilocular structures (Fig. 20) elongated and cone-shaped, either to the side or at ends of branches, sometimes hair-tipped 21

20. **Plastids disc-shaped** (drawing below) 23

cell with band-shaped plastids

cell with disc-shaped plastids

21. **Filaments heavily entangled, forming woolly and spongy rope-like strands; hook-like branchlets present;** dark brown tufts, up to 15 cm long; plurilocular structures often visible; epiphytic (mostly on fucoids); uncommon; found from late winter to late fall, on the open coast, in the intertidal zone .... *Spongonema tomentosum* (Fig. 19)

21. **Filaments not heavily entangled; hook-like branches absent; tufts very soft, not spongy** 22

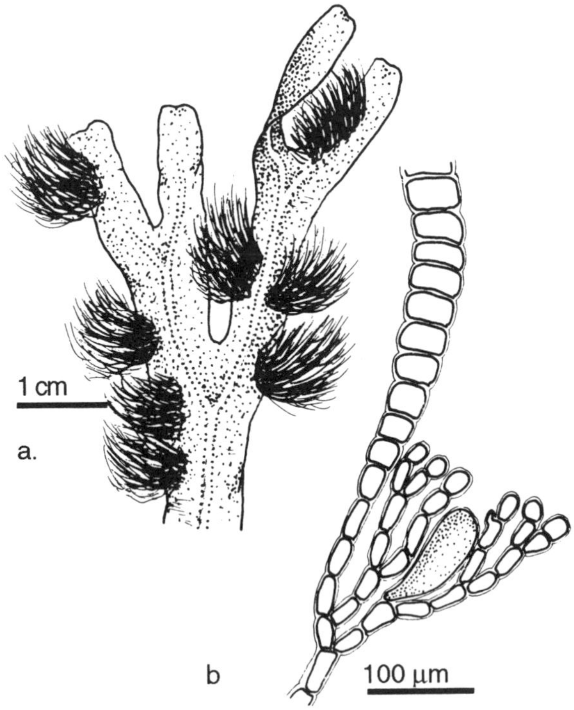

Fig. 17. ***Elachista fucicola***
a. tufts growing on *Fucus* sp.;
b. uniseriate filament, branched at base only , and with unilocular sporangium

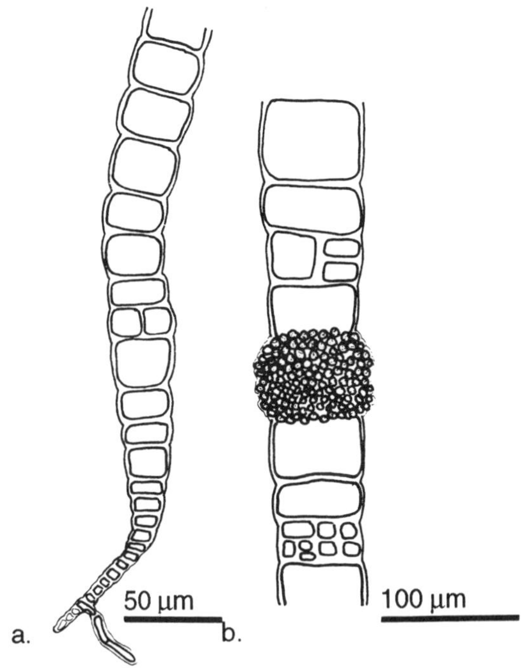

Fig. 18. ***Halothrix lumbricalis***
a. single, uniseriate filament
b. filament with plurilocular sporangia

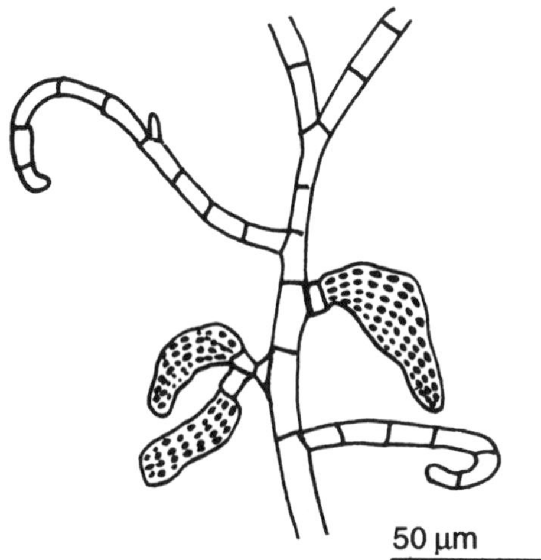

Fig. 19. ***Spongonema tomentosum***
portion of branch with hook-like branchlets, and irregular, recurved, plurilocular sporangia

22. **Abundant fascicles present (short clusters of branches arising from the same point); primary filaments with few branched below and often twisted together;** brown tufts, up to 20 cm long; cells rectangular near base, slightly barrel-shaped in upper parts; plurilocular structures often present; found year round, but uncommon; generally epiphytic on blades of Laminariales, but also attached to rocks, on the open coast and in estuaries .... *Ectocarpus fasciculatus* (Fig. 20)

22. **Fascicles absent; branching abundantly from base;** tufts yellowish brown, up to 50 (75) cm long; cells generally rectangular; terminal, plurilocular structures often present; very common year round, on the open coast and in estuaries, in the intertidal and subtidal zones, down to -20 m, attached to rocks, or epiphytic; often found free floating .... *Ectocarpus siliculosus* (Fig. 21)

23. **Reproductive structures intercalary (in a row, within the filament) as well as terminal; with occasional opposite branching; filaments often twisted and forming visible, rope-like stands;** very common year round, on the open coast and in estuaries; found in the intertidal and shallow subtidal zones, usually epiphytic on coarser algae, but also attached to rocks; often free floating ....*Pilayella littoralis* (Fig. 22)

23. **Reproductive structures lateral (to the side of an axis) or terminal (at the tip of a branch), generally oval and rounded** 24

24. **Reproductive structures in clusters;** found year round, attached to larger algae or in the drift; uncommon .... *Botrytella micromora* (Fig. 23)

24. **Reproductive structures either single, or in pairs and opposite, or sometimes in rows on the side of a branch** 25

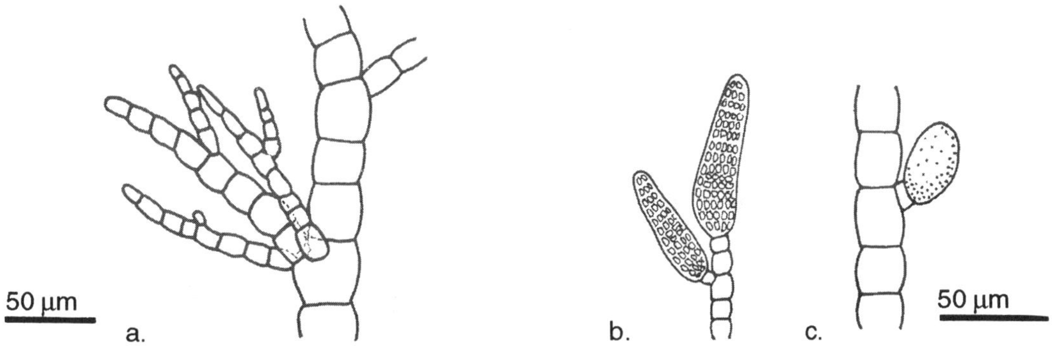

Fig. 20. ***Ectocarpus fasciculatus***; a. portion of branch with fascicle; b. plurilocular sporangia; c. unilocular sporangium

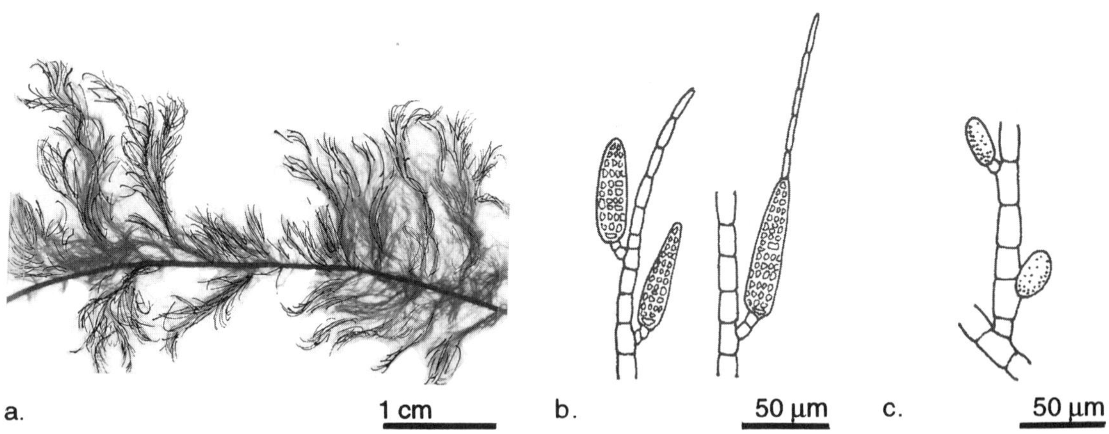

Fig. 21. ***Ectocarpus siliculosus;*** a. tufts on host; b. plurilocular sporangia; c. unilocular sporangia

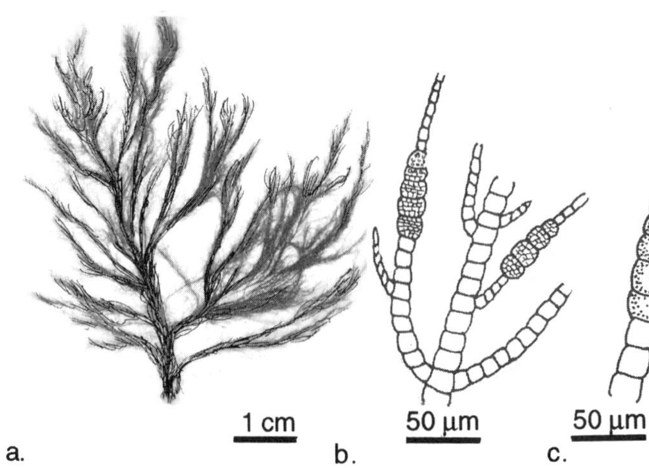

Fig. 22. ***Pilayella littoralis***; a. tuft; b. branching pattern and plurilocular sporangia; c. unilocular sporangia

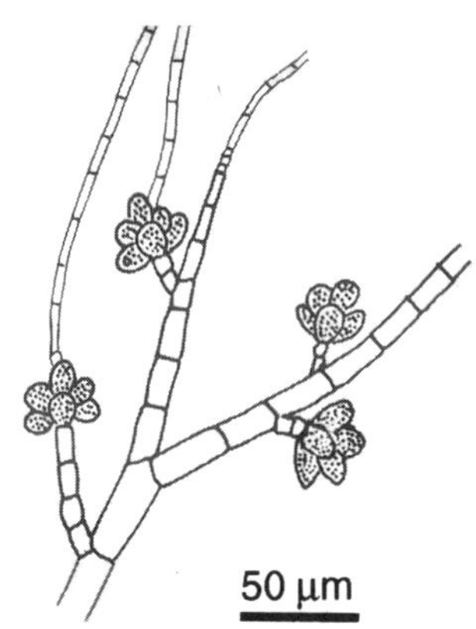

Fig. 23. ***Botrytella micromora*** branching pattern and unilocular sporangia in clusters

25. **Plurilocular sporangia cylindrical; branching irregular; tips of branches often ending in a long, thin, hair-like filament;** uncommon north of Cape Cod, but more abundant to the south; found on the open coast and in estuaries, attached to rocks, stones, shells, or epiphytic on other algae; (♦1 of 4) .... *Hincksia mitchelliae* (Fig. 24)

25. **Plurilocular sporangia cone-shaped or oval; branching often opposite** 26

26. **Axes and branches uniseriate throughout;** common year round, on the open coast and in estuaries, in the intertidal and subtidal zones, down to -12 m, generally epiphytic on other algae; (♦1 of 4) .... *Hincksia granulosa* (Fig. 25)

26. **Longitudinal cell division occurring in older filaments, forming a pluriseriate structure; lateral unilocular sporangia rounded, often in pairs and opposite;** tufts no more than 5 cm long; uncommon; found on the open coast and in estuaries, in spring and summer, attached to shells or epiphytic on other algae .... *Isthmoplea sphaerophora* (Fig. 26)

27. **Thallus an irregular, hollow, globular mass with lumps**; pale brown to yellow when young, turning dark brown with age; found in the intertidal zone or just below, on the open coast, usually epiphytic on other algae; growing from early spring to summer, south of Cape Cod, but extending into fall to the north .... *Leathesia difformis* (Fig. 27)

27. **Thallus otherwise**.. 28

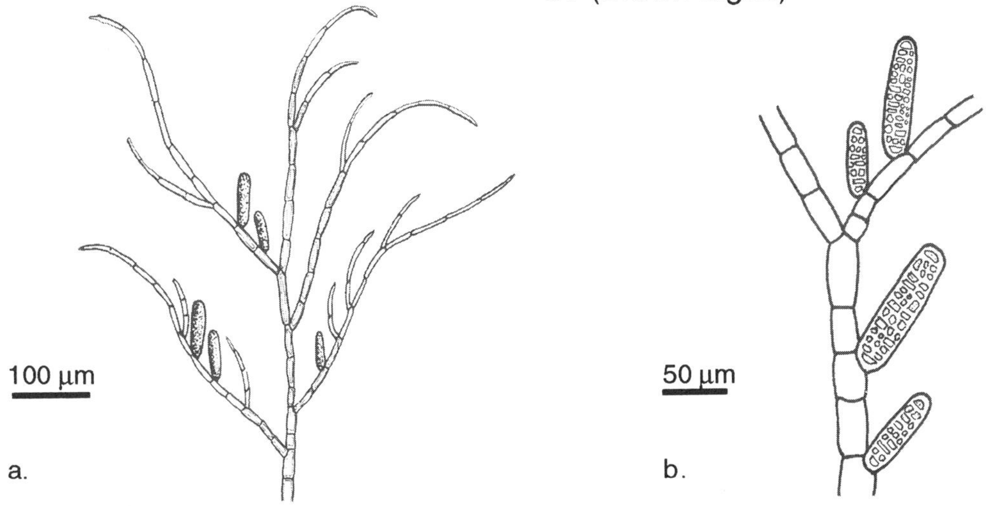

Fig. 24. ***Hincksia mitchelliae***; a. uniseriate filament showing long, hair-like tip of branches; b. cylindrical plurilocular sporangia

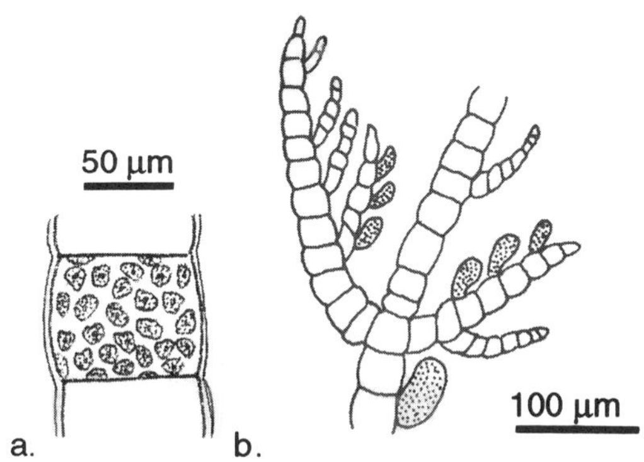

Fig. 25. ***Hincksia granulosa***
a. disc-shaped chromatophores;
b. uniseriate filaments with opposite branching and oval plurilocular sporangia

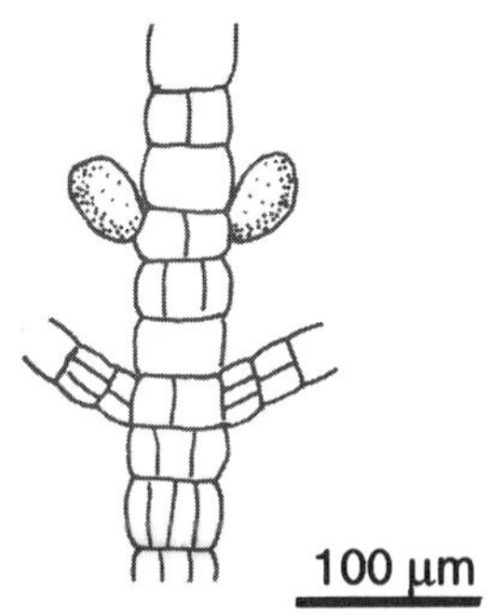

Fig. 26. ***Isthmoplea sphaerophora*** pluriseriate filament with rounded, opposite unilocular sporangia

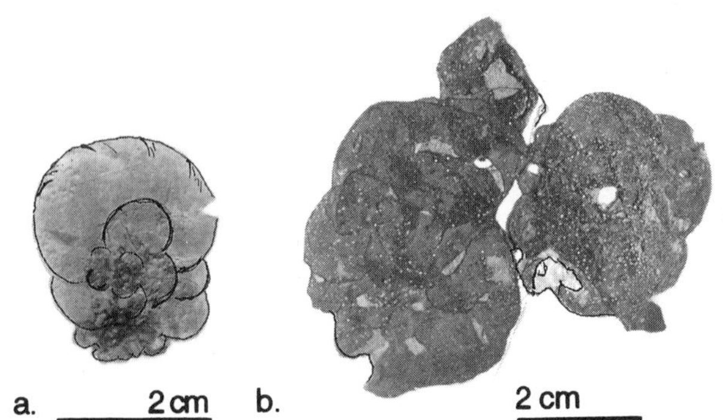

Fig. 27. ***Leathesia difformis***; a. young, individual specimen (spring form); b. older, mature individual (summer form)

28. **Unbranched tube, cord or blade; blade may be split lengthwise into finger-like straps** 29
28. **Thallus distinctly branched** 43

29. **Thallus a cylindrical tube or cord; constrictions may be present** 30
29. **Thallus a flat blade, with or without a midrib, holes or hairs** 33

30. **Thallus cord-like, solid and firm, but slippery**, up to 1 m (8 m) long 31
30. **Thallus a hollow tube, not particularly slippery**, up to 50 cm long 32

31. **Thallus cord-like, and covered with a sheath of pale brown (yellowish or green) hairs**; hairs of uniform length (0.5-2 cm); found from spring to fall, on the open coast and in estuaries, in the subtidal zone, down to -12 m, attached to rocks.......... *Chorda tomentosa* (Fig. 28)
31. **Thallus not covered with pigmented hairs at maturity** (young specimens may have clear, deciduous hairs); found year round, on the open coast and in estuaries, in the subtidal zone, down to -10 m, attached to rocks, often growing in clumps; common, particularly south of Cape Cod .... *Chorda filum* (Fig. 29)

32. **Sausage-like constrictions usually present in mature individuals; unconstricted forms also found as long, narrow, light brown tubes tapering to base;** both forms are common year round, on the open coast and in estuaries, around the low tide level and just below, attached to rocks, or epiphytic on other algae; often growing in clumps or large patches; also common in tide pools .... *Scytosiphon simplicissimus* (Fig. 30)
32. **Thallus less than 5 mm in diameter, sometimes contorted; tufts of reproductive structures and hairs producing a rough, spotted surface**; uncommon; found in late winter to midsummer, on the open coast, around the low tide level, attached to rocks, and epiphytic on other algae .... *Asperococcus fistulosus* (Fig. 31)

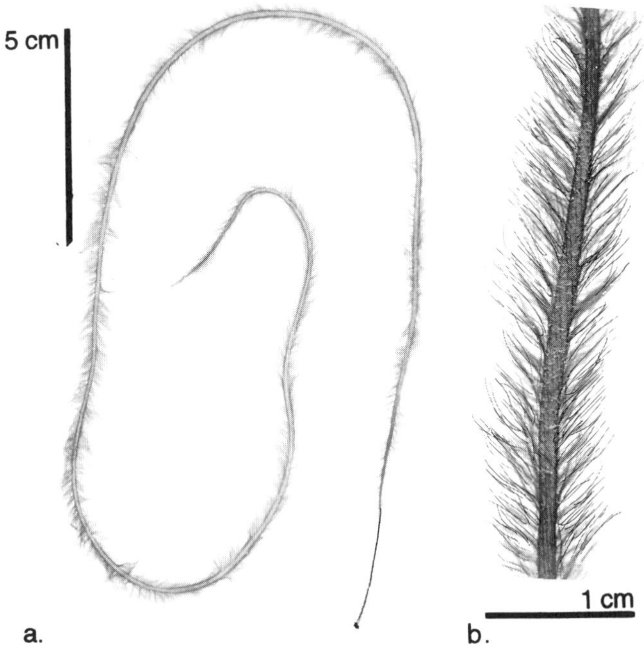

a. b.

Fig. 28. ***Chorda tomentosa***
a. individual specimen; b. detail of hairs

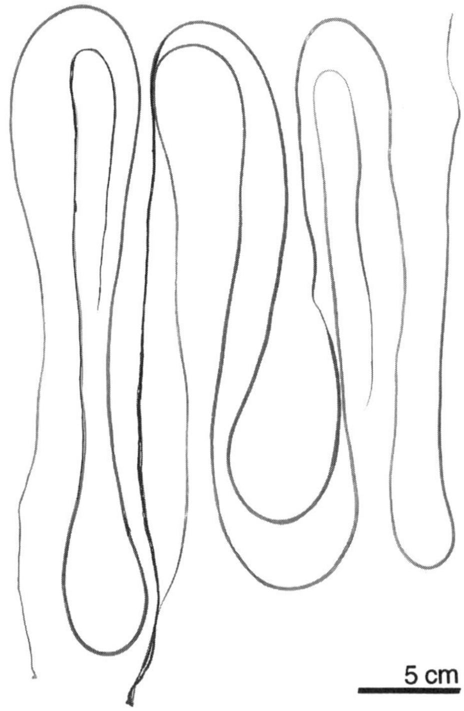

Fig. 29. ***Chorda filum***
individual specimens

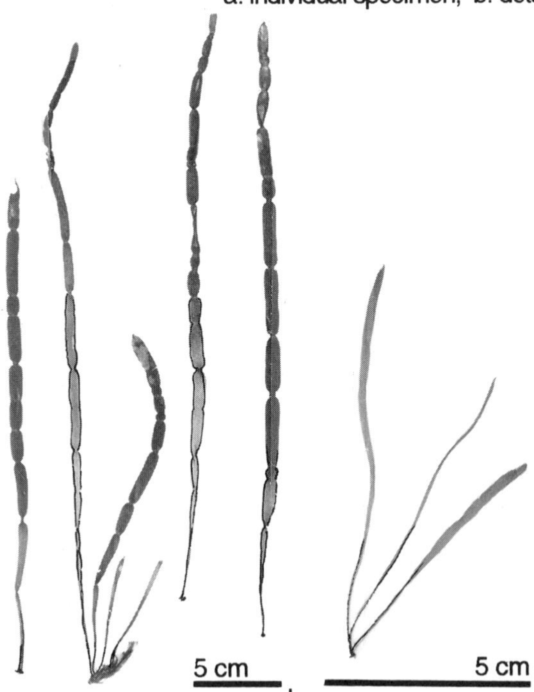

a. b.

Fig. 30. ***Scytosiphon simplicissimus***
a. specimens with constrictions;
b. specimens lacking constrictions

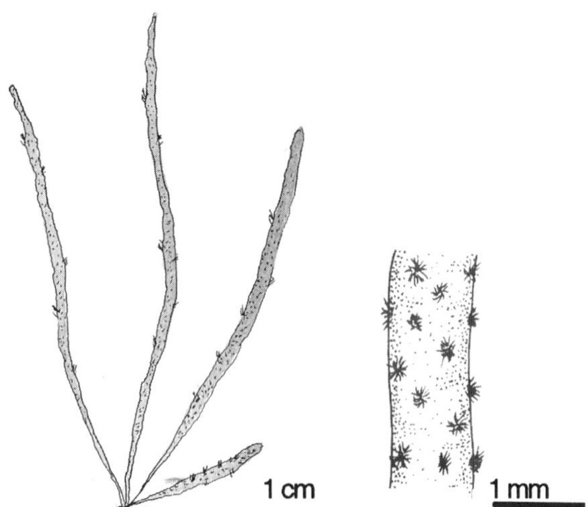

a. b.

Fig. 31. ***Asperococcus fistulosus***
a. cluster of individual specimens showing spotted surface;
b. detail of surface with tufts of hairs

33. **Holdfast well differentiated, disc-like, or branched and root-like; stipe cylindrical or flattened; blade flat and leathery,** often 2 m long or more 34

33. **Thallus consisting mostly of a strap or ribbon-shaped blade,** up to 50 cm long, by no more than 5 cm wide; holdfast a very small disc; stipe inconspicuous; microscopic characteristics are needed to identify the following taxa 39

34. **Midrib present the length of the blade** (well marked central strip, thicker than the blade) 35

34. **Midrib absent** 36

35. **Blade coarse, undivided, large, with numerous holes and a midrib;** up to 2 m long, by up to 60 cm wide; light to dark brown; short, stout stipe and thick holdfast; common year round, from Cape Cod northward, on the open coast, in deep waters, from about -10 to -32 m; attached to rocks; often found in the drift .... *Agarum clathratum* (Fig. 32)

35. **Blade without holes, long and narrow, sometimes ruffled; midrib present; sporophylls (short leafy growths) present on stipe, when reproductive;** up to 2 m long or more; blade often torn to the midrib; yellowish to greenish brown; common year round, from Cape Cod northward, in exposed areas on the open coast, from the low tide level, down to -18 m, attached to rocks; often found in the drift .... *Alaria esculenta* (Fig. 33)

36. **Blade surface with scattered, small tufts of hairs; upper stipe flattened; large, disc-shaped holdfast;** blade simple when young, becoming divided; up to .75 m wide, by up to 2 m long; brown to yellow brown; found in the summer, from Cape Cod northward, in exposed areas, on the open coast, attached to rocks, in the subtidal zone, down to -30 m or more; also found attached in low tide pools, or in the drift .... *Saccorhiza dermatodea* (Fig. 34)

36. **Tufts of hairs absent from surface of blade;** holdfast root-like; when reproductive, sporangia form dark brown patches or large, irregular strips, in the center of the blade 37

Fig. 32. ***Agarum clathratum***
individual specimen with midrib and perforations

Fig. 33. ***Alaria esculenta***
a. young specimen with midrib
b. mature specimen with midrib and sporophylls

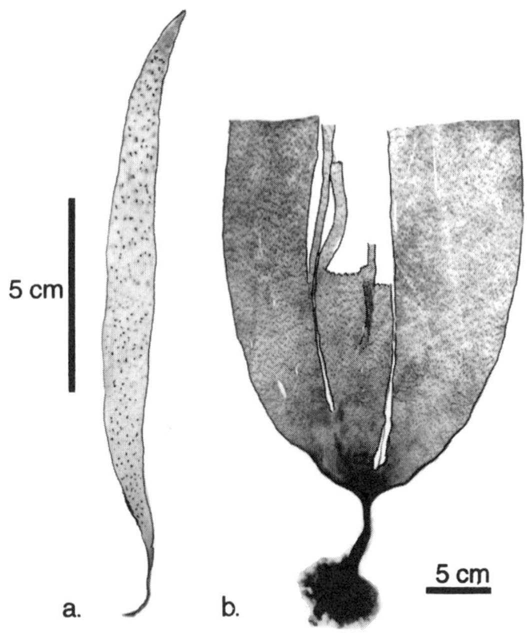

Fig. 34. ***Saccorhiza dermatodea***
a. young specimen dotted with tufts of hairs;
b. mature specimen with divided blade, tufts of hairs, and large holdfast

37. **Broad blade, divided longitudinally into strap-like segments**; up to 15 m long; stipe thick and short; holdfast heavy; common year round, in very exposed areas, on the open coast, in the subtidal zone, down to at least -18 m, attached to rocks and shells; often found in the drift, after a storm .... *Laminaria digitata* (Fig 35)

37. **Blade a single, undivided strap; edges sometimes undulated and ruffled** 38

38. **Stipe short and solid throughout**; very common year round, on the open coast and in estuaries; attached to rocks and shells, in the subtidal zone, down to at least -30 m; often in the drift, after a storm ....*Laminaria saccharina* (Fig. 36)

38. **Stipe hollow and swollen in upper parts, often longer than blade, up to 2 m long or more;** blade 2-7 (12) m long, by 50-70 cm wide; found year round, in the subtidal zone, down to at least -30 m; attached to rocks, on the open coast and in estuaries; more common from Cape Cod northward. This taxon may not be a true species and may be included in *L. saccharina* (South and Tittley, 1986) .... *Laminaria longicruris* (Fig. 37)

39. **Seven or more layers of cells in cross-section of blade (140-200 μm), with cells of 2 distinctly different sizes** (larger, clear medullary cells in center and smaller, pigmented cortical cells on both surfaces); hairs not in tufts 40

39. **Two to 7 cells thick in cross-section of blade, with cells of approximately equal size across, or only slightly different** 41

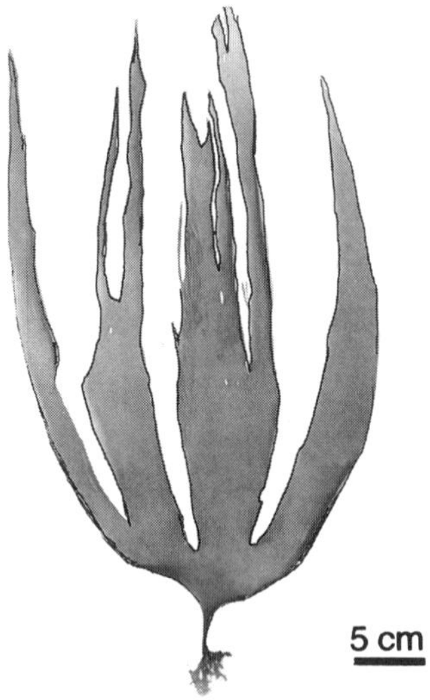

Fig. 35. ***Laminaria digitata***
individual specimen with
a divided blade

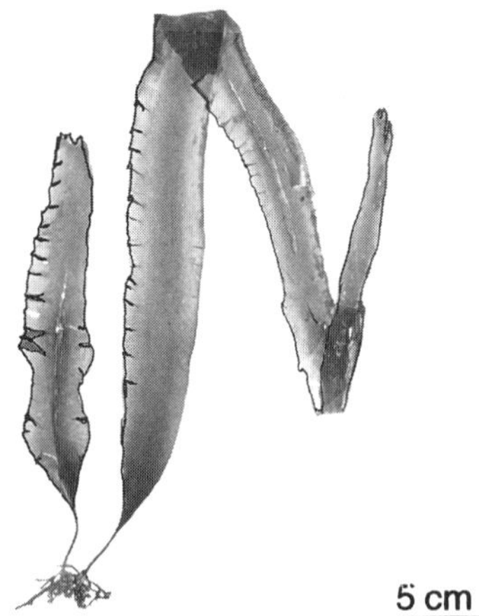

Fig. 36. ***Laminaria saccharina***
individual specimens

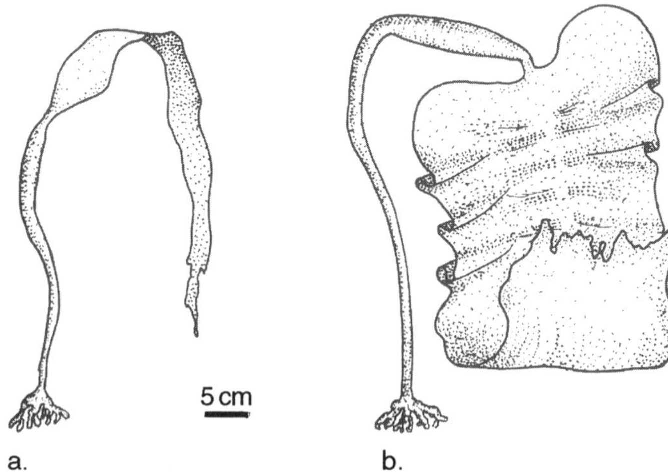

Fig. 37. ***Laminaria longicruris***

a. young specimen with somewhat inflated stipe

b. mature specimen with ruffled blade and distinctly swollen stipe

40. **Blade up to 20 cm long, by up to 4 cm wide**; base usually asymmetrical; common from late fall to early summer south of Cape Cod, but year round to the north; found on the open coast and in estuaries, attached to rocks, or epiphytic on other algae, in the intertidal and subtidal zones, down to -12 m; often found in tide pools .... *Petalonia fascia* (Fig. 38)

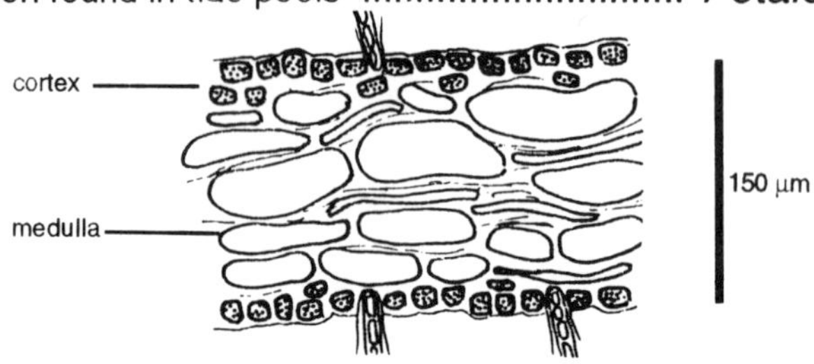

P. fascia,
cross-section

40. **Blade narrow, .5 to 2 (4) mm wide, by 15 (23) cm long;** often growing in clumps; uncommon; found from spring to early winter, on the open coast and in estuaries, attached to rocks, around the low tide level... *Petalonia zosterifolia* (Fig. 39)

41. **Two (rarely 3 or 4) cells in cross-section (20-40 µm); blade usually less than 5 cm long**; pale brown, sometimes greenish; rows of hairs at edges of blade when young; a spring species south of Cape Cod, it survives through the summer to the north; epiphytic on eelgrass and other algae, on the open coast and in estuaries, in the low intertidal and subtidal zones, down to -3 m; also found in tide pools .. .. *Punctaria tenuissima* (Fig. 40)

P. tenuissima
cross-section

41. **Four (rarely 2) to 7 cells in cross-section; blade up to 50 cm long by 3 (5) cm wide, with tufts of hairs on entire surface producing a spotted or mottled appearance;** tips often becoming shredded and covered with epiphytes 42

42. **Blade 2 to 4 cells thick (50-85 µm),** (drawing below); 10-45 cm long by 2-8 (15) cm wide; light brown, sometimes greenish; texture soft, papery ; uncommon; found from midwinter to fall, south of Cape Cod, but only in summer to the north; growing on the open coast and in estuaries, in the low intertidal and subtidal zones, down to -12 m, attached to rocks, or epiphytic on other algae and eelgrass .... *Punctaria latifolia* (Fig. 41)

42. **Blade 4 to 7 cells thick (100-225 µm),** (drawing below); up to 20 cm long or longer; texture firm, leathery; uncommon; found from late winter to summer, on the open coast and in estuaries, around the low tide level, down to -6 m, attached to rocks, or epiphytic on other algae, and on eelgrass .... *Punctaria plantaginea* (Fig. 42)

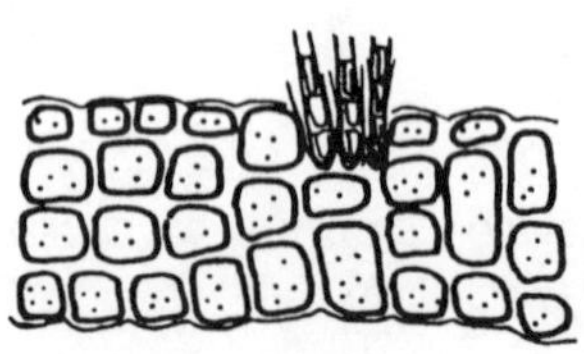

50 µm

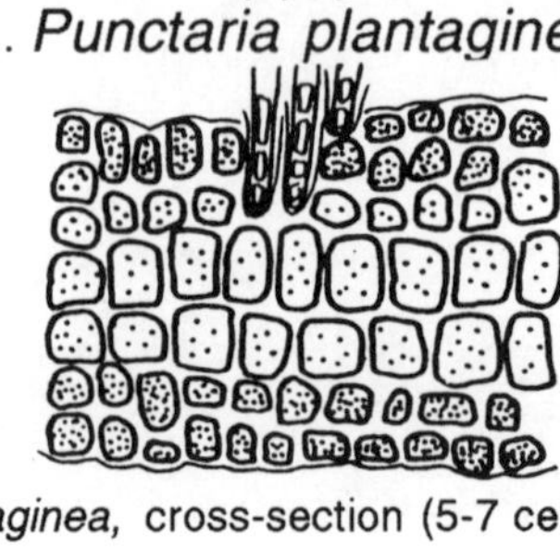

P. latifolia, cross-section (4 cells thick)

P. plantaginea, cross-section (5-7 cells thick)

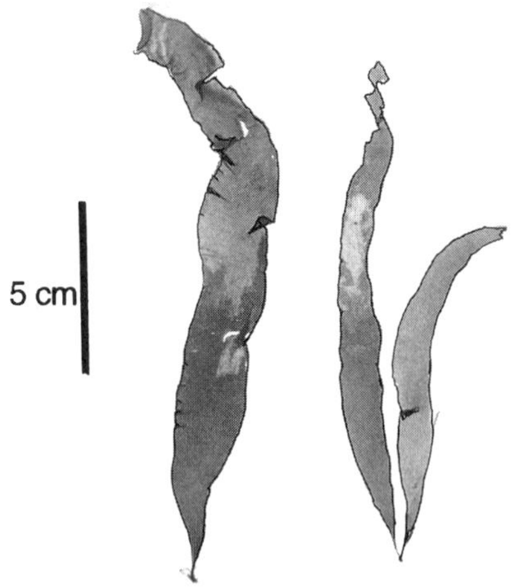

Fig. 39. ***Petalonia fascia***
individual specimens

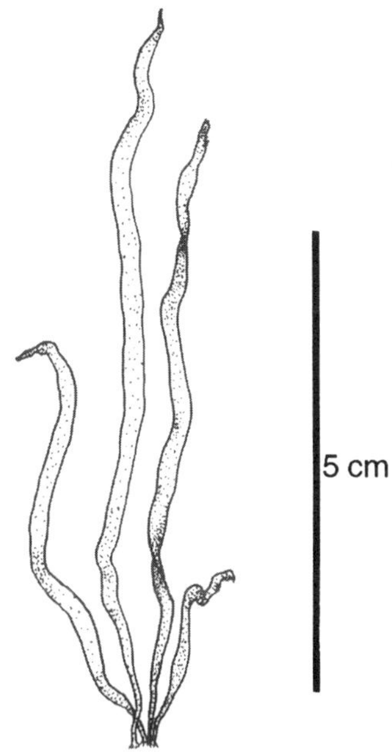

Fig. 39. ***Petalonia zosterifolia***
cluster of individual specimens

Fig. 40. ***Punctaria tenuissima***
blades attached to host

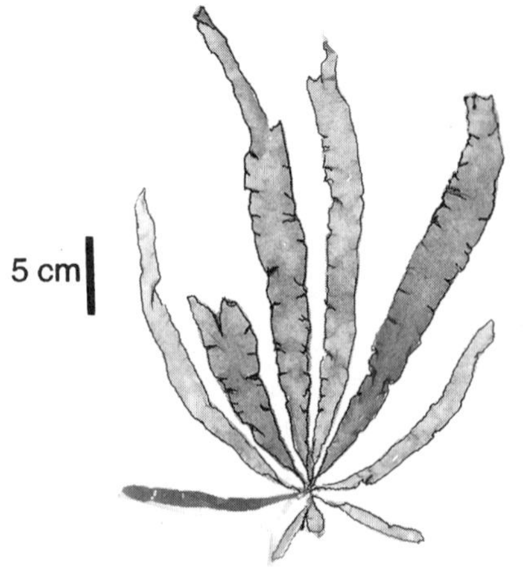

Fig. 41. ***Punctaria latifolia***
clump of individual specimens

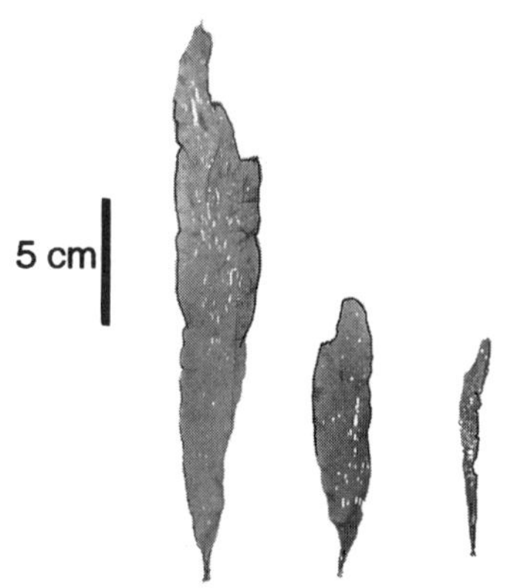

Fig. 42. ***Punctaria plantaginea***
individual specimens

43. **Axes, branches and appendages clearly compressed or flattened,** 3 mm or wider in flat areas 44

43. **Thallus tubular or not conspicuously flattened, at least at base;** 3 mm or less in widest areas, branches in some species becoming very thin and hair-like; some species profusely branched, producing a bushy, filamentous appearance in the upper parts of the thallus.. 50

44. **Elongated, leaf-like appendages growing from a branched, stem-like axis;** small, round vesicles present; found year round, from Cape Cod southward, on the open coast, but more common in estuaries, in the subtidal zone, down to -30 m, attached to rocks and shells; also found in the drift; (♦1 of 3)* .... *Sargassum filipendula* (Fig. 43)

44. **Leaf-like appendages absent; tips of branches forming receptacles** (swollen in some species, drawing below) bearing male and/or female gametes (drawing below) 45

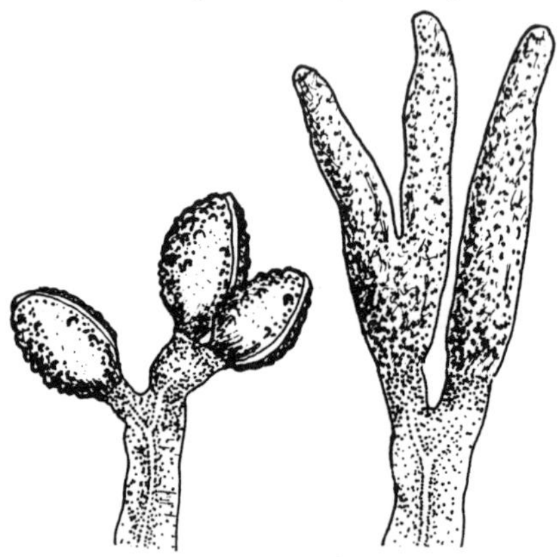

Fucus spiralis *Fucus distichus* ssp. *edentatus*

receptacles

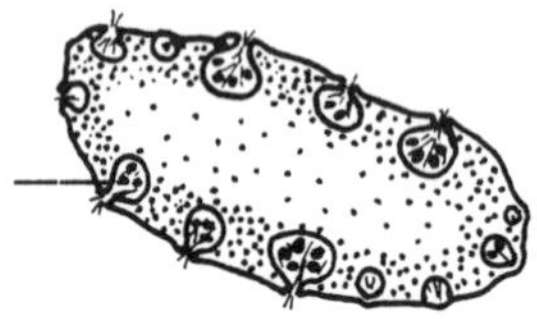

cross section of receptacle

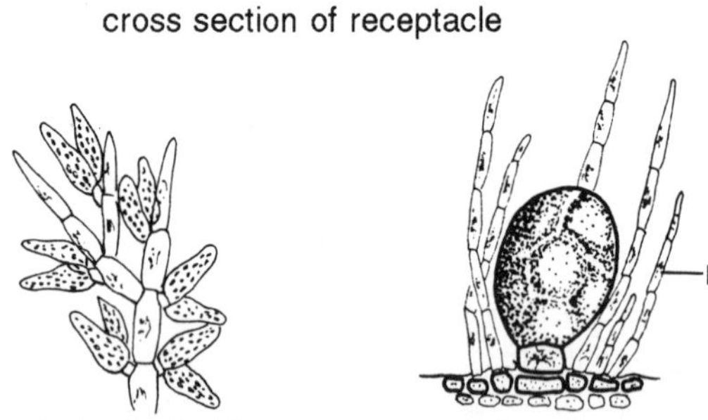

cluster of antheridia oogonium with eggs

sexual structures

45. **Branching irregular, major axes bearing short side branches, midrib absent;** single vesicles present along main axes; lateral, swollen receptacles present when reproductive; thallus dark brown but sometimes yellow or greenish; up to1 m in length or more; very common year round, in semi-exposed and sheltered areas of the open coast and estuaries, attached to rocks, in the mid to low intertidal zone. In salt marshes and tidal flats, free floating specimens display a modified morphology; branching becomes very irregular, entangled, and vesicles are few or absent (*A. nodosum* ecad *scorpioides*) .... *Ascophyllum nodosum* (Fig. 44)

45. **Branching dichotomous** (Y-shaped and even on both sides); **midrib flanked by two flat wings;** wings often torn or disappearing at base 46

* The other 2 species of *Sargassum* recorded for this area can also be found in the drift, south of Cape Cod. They are brought in from the Sargasso Sea and the Caribbean by currents and storms. They are both sterile and lack a holdfast. *Sargassum natans* has vesicles tipped with a small spine. *Sargassum fluitans* has vesicles without spines.

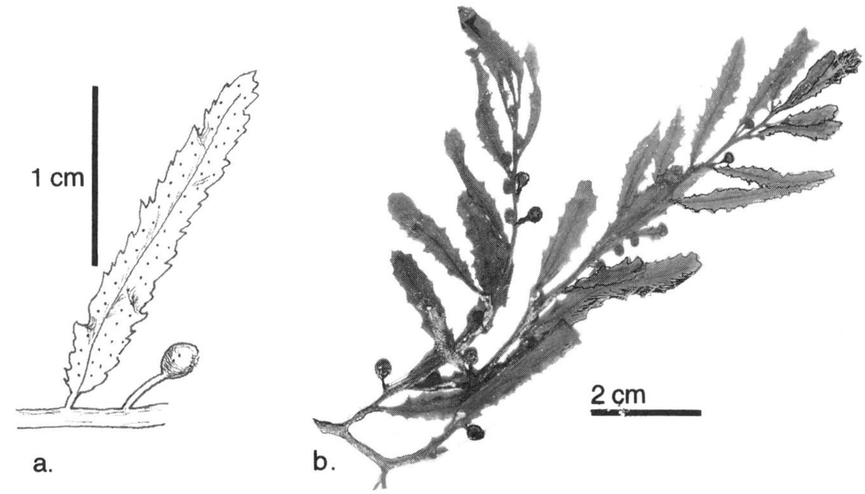

Fig. 43. ***Sargassum filipendula***
a. "leafy" blade with vesicle;
b. branches with vesicles and blades

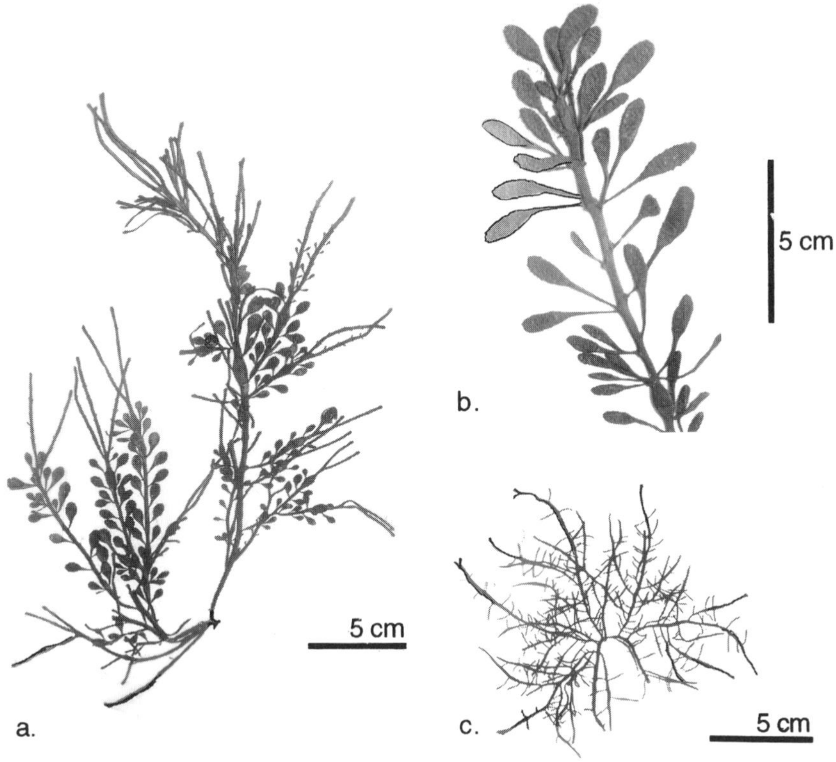

Fig. 44. ***Ascophyllum nodosum***
a. typical mature specimen; b. tip of branch with mature receptacles
c. individual specimen, *A. nodosum* ecad *scorpioides*

46. **Paired vesicles on wings (see below); male and female on separate individuals** 47
46. **No vesicle on wings; sexes on same individual, in same conceptacles** (drawing, step 44) 48

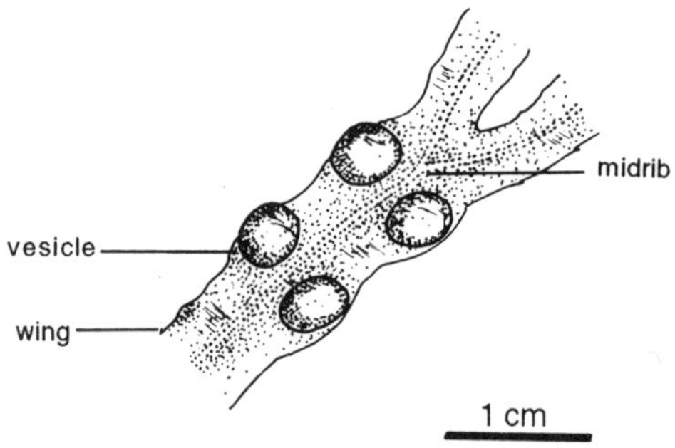

F. vesiculosus,
portion of branch showing midrib
and paired vesicles, or air bladders

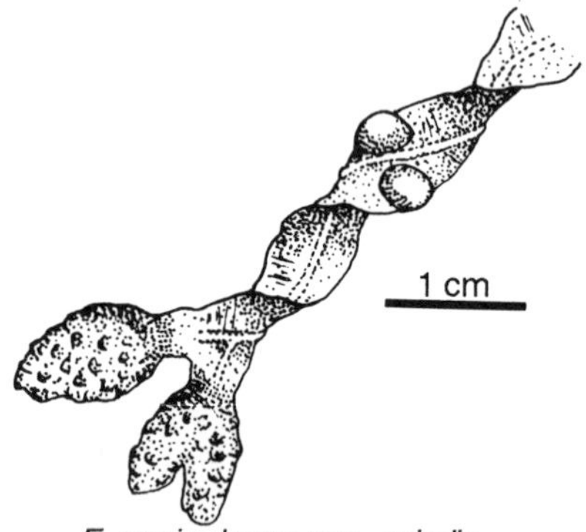

F. vesiculosus var. *spiralis*
portion of branch showing twisting

47. **Axes and branches strap shaped, not regularly twisted, pairs of vesicles present**; large, golden to dark brown thallus, up to 70 cm or more and up to 1.5 cm wide; found year round attached to rocks, on the open coast and in estuaries, in the mid to low intertidal zone and just below; vesicles may be absent in certain exposed conditions and examining the gametes with the microscope may be the only way to separate this species from others. One of the most common species in New England .... *Fucus vesiculosus* (Fig. 45)
47. **Thallus similar to *Fucus vesiculosus,* but axes and branches spirally twisted** (see above); common year round in estuaries; attached to rocks and entangled at the base of plants, in tidal flats... *Fucus vesiculosus* var. *spiralis*

48. **Receptacles swollen, round or oval, usually with a ridged edge** (see drawing, step 44); thallus up to 25 cm long; very common year round, attached to rocks, in the upper intertidal zone where it forms a narrow band along with barnacles .... *Fucus spiralis* (Fig. 46)
48. **Receptacles flattened or barely swollen at maturity*** 49

49. **Receptacles long and pointed, length 4-10 times width, up to 15 cm long**, sometimes forked (see drawing, step 44); large thallus, up to 75 cm long or more; found year round, attached to rocks, in exposed areas, on the open coast, in the intertidal zone and just below... *Fucus distichus* ssp. *edentatus* * (Fig. 47)
49. **Receptacles short and blunt, length 1-2 times width, edges with a slight wing or ridge;** thallus up to 70 cm long; found year round, attached to rocks, in sheltered areas, in the intertidal and just below; generally growing in estuaries, but occasionally on the open coast .... *Fucus distichus* ssp. *evanescens** (Fig. 48)

* South and Tittley (1986) as proposed by Rice and Chapman (1985), have regrouped the two following subspecies under the name *Fucus evanescens.* In New England, the two subspecies are very distinct and they are maintained as separate entities for the purpose of this key.

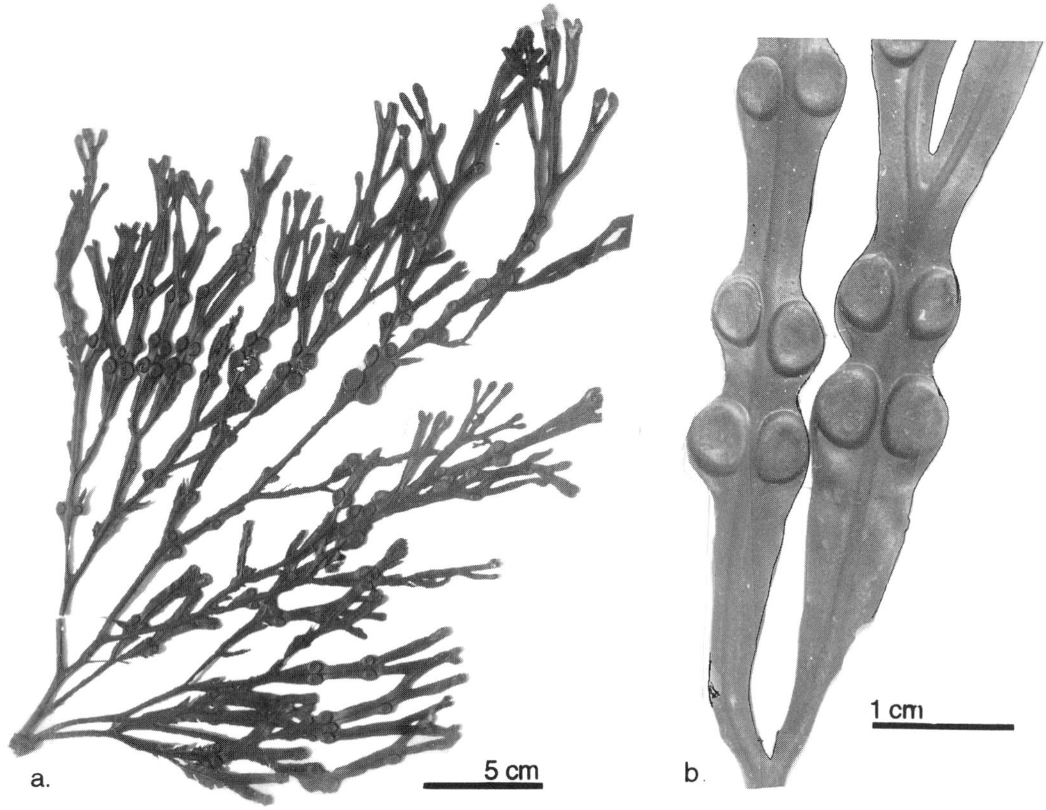

Fig. 45. ***Fucus vesiculosus***
a. individual specimen with dichotomous branching
b. detail from a pressed specimen, showing midrib and vesicles

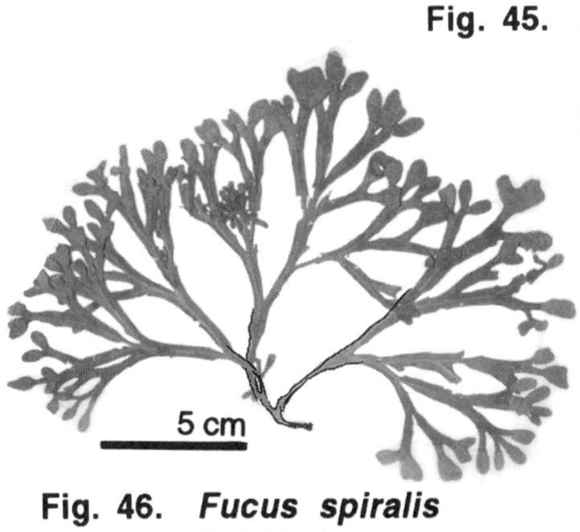

Fig. 46. ***Fucus spiralis***
individual specimen

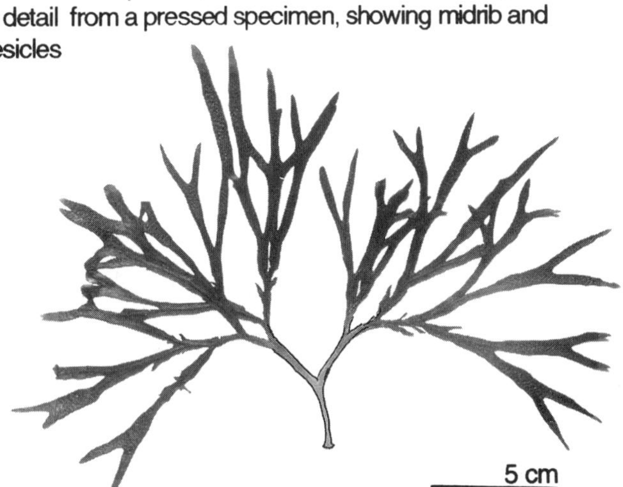

Fig. 47. ***Fucus distichus*** **ssp.** ***edentatus***
individual specimen

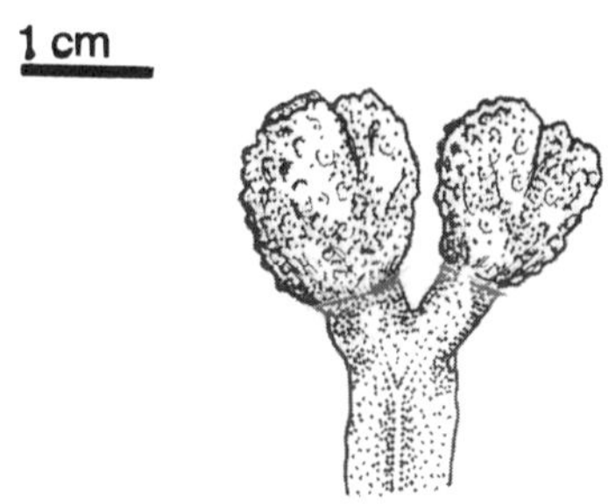

Fig. 48. ***Fucus distichus*** **ssp.** ***evanescens***
tip of branch with flattened receptacles

50: **Thallus pale brown to light olive-brown, worm-like, very soft, gelatinous and slippery**; branching often at right angles; loosely organized axes with interwoven filaments, and a perpendicular sheath of short, photosynthetic filaments (see below), easily spread into individual filaments by crushing between slide and cover slip 51

portion of axis with sporangium and filaments at right angles

50. **Not soft and worm-like**; thallus may be somewhat slippery and gelatinous, but filaments can not be spread by crushing (solid construction) 52

51. **Thick axes, (1) 3-5 (10) mm in diameter; colorless hairs without a basal sheath;** thallus up to 30 cm in length, with a few long branches and shorter branchlets; found at low tide level or just below, on the open coast and in estuaries, from spring to fall, in warm waters, attached to rocks, shells or epiphytic on other algae and at the base of eelgrass.................... *Eudesme virescens* (Fig. 49)

51. **Smaller thallus, 2 mm in diameter or less, often contorted or curly; with a sheath at the base of colorless hairs**; axes up to 30 cm long, usually thin and slender, with few branches; epiphytic mostly on eelgrass where it forms dense clusters, but also found on other algae, in spring and summer, on the open coast and in estuaries .... *Cladosiphon zosterae* (Fig. 50)

52. **Short tufts, 5 cm long or less; growth from large cell, at tip of filament** .... *Sphacelaria* spp. (p. 42, step 12)

52. **Thallus 5 cm and longer** 53

53. **Axes and branches covered with a densely packed sheath of short, whorled tufts** (resembling a branched, dark brown pipe cleaner); often stiff, with wiry branches, up to 30 cm in height; base of thallus often denuded and wiry; found year round, south of Cape Cod; rare to the north; attached to rocks, in the subtidal zone, on the open coast and in estuaries; also found in the drift after storms .... *Cladostephus spongiosus* (Fig. 51)

53. **Thallus otherwise** 54

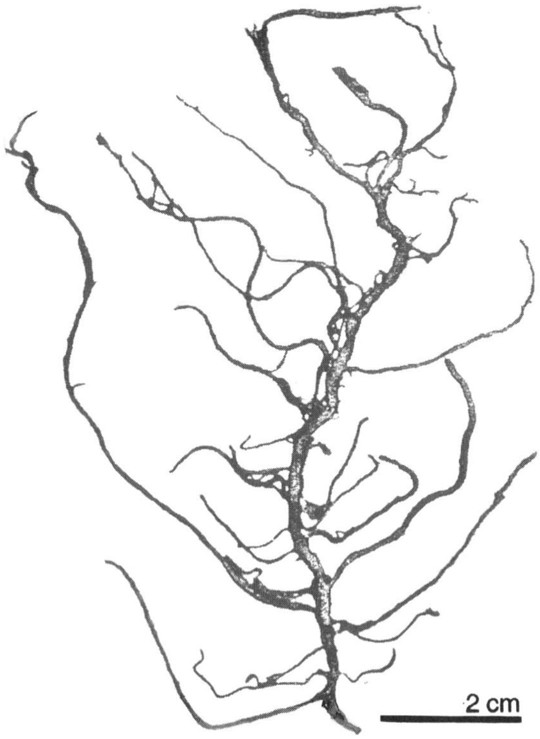

Fig. 49. ***Eudesme virescens***
portion of specimen

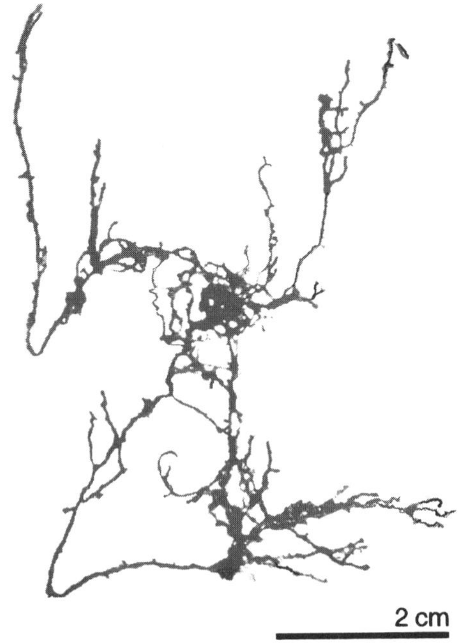

Fig. 50. ***Cladosiphon zosterae***
portion of specimen

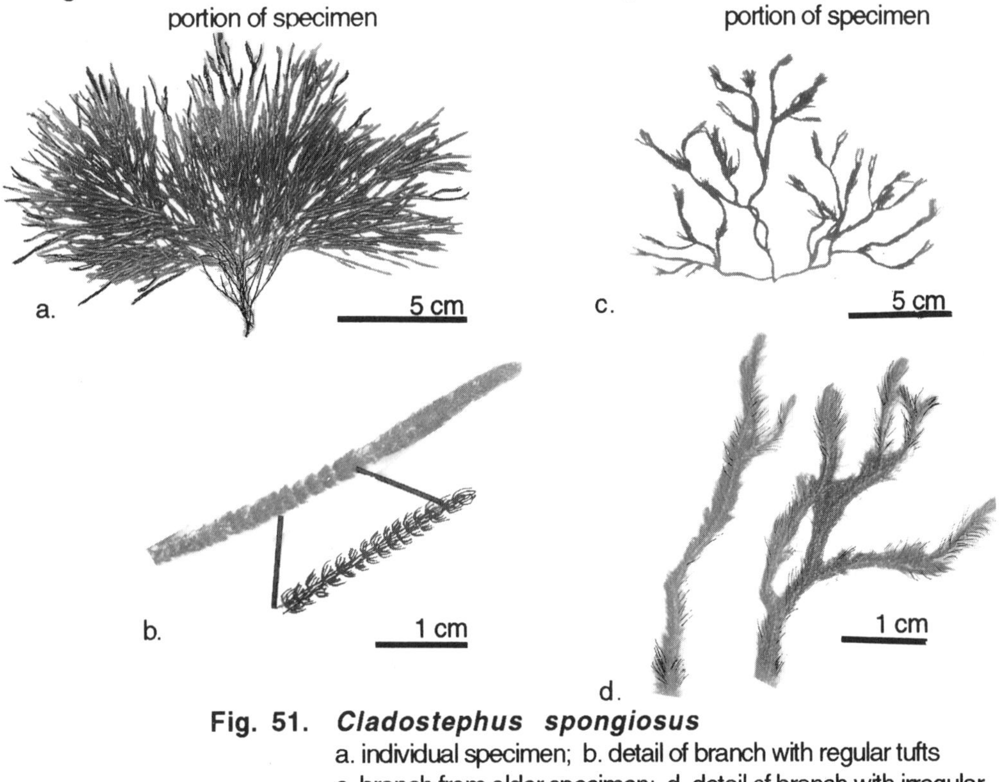

Fig. 51. ***Cladostephus spongiosus***
a. individual specimen; b. detail of branch with regular tufts
c. branch from older specimen; d. detail of branch with irregular tufts

54. **Covered with fine hairs, at least when young;** large, bushy thallus, highly branched; delicate in spring, becoming coarse in summer; often turning green when dried 55
54. **Hairs absent** 57

55. **Some branches narrowly flattened, with short, alternate teeth pointing forward at regular intervals** (see below); covered with brown to greenish tufts of hairs in spring, in opposite, occasionally alternate pairs; hairs 1-3 mm long, and 30-50 µm wide; thallus large, up to 2 m, somewhat stiff, dark brown, with abundant branching; common year round, with young specimens developing in late winter; found on the open coast and in estuaries, in the subtidal zone, down to -26 m, attached to rocks; often free floating; specimens surviving through winter, often have new regrowth at old branching forks in the spring * .... *Desmarestia aculeata* (Fig. 52)

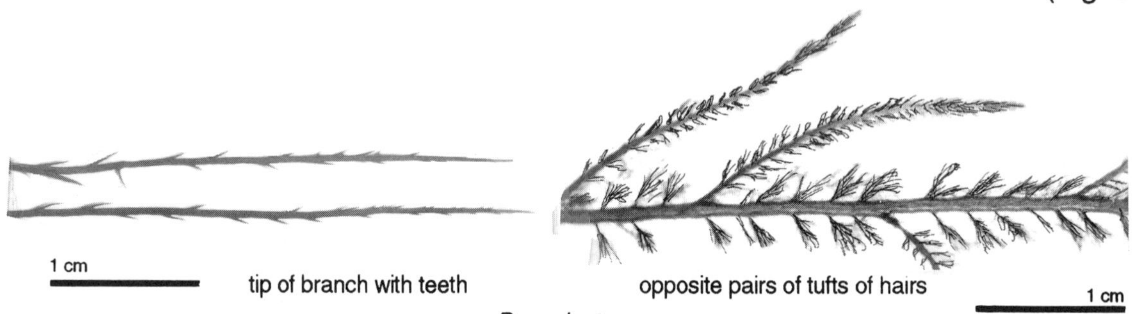

D. aculeata

55. **Branches remaining cylindrical with no teeth** 56

56. **Branching opposite; ultimate branches ending in fine hairs which are shed through the spring; thallus soft and bushy;** light brown to olive-green, abundantly branched, up to 60 cm long; developing in late winter and disappearing by late fall; found on the open coast and in estuaries, in low tide pools and in the subtidal zone, down to -20 m, attached to rocks; often free floating * .... *Desmarestia viridis* (Fig. 53)
56. **Branching irregular (including opposite, alternate or whorled); covered with persistent, whorled tufts of hairs**; hairs light brown to green, up to 4 mm long, by 7-21 µm wide; thallus stiff, dark brown, up to 40 cm long; uncommon; found from Cape Cod southward only, from late winter to late fall, in the subtidal zone, down to -9 m, attached to stones and shells; often free floating .... *Arthrocladia villosa* (Fig. 54)

57. **No distinguishable main axis;** branching irregular; branches at tips often at right angles 58
57. **Main axis clearly distinguishable** 59

* *Desmarestia* species have highly acidic vacuoles containing sulfuric and malic acids (pH 0.8 to 1.8). As soon as these organisms are removed from the water, the vacuole membrane (tonoplast) is destroyed and the cells are killed. A healthy, brown thallus will rapidly turn green and decay quickly in a bucket. In addition, specimens of *Desmarestia* spp. could damage other species if left in contact with them. They should be isolated in plastic bags and kept chilled or pressed as soon as possible.

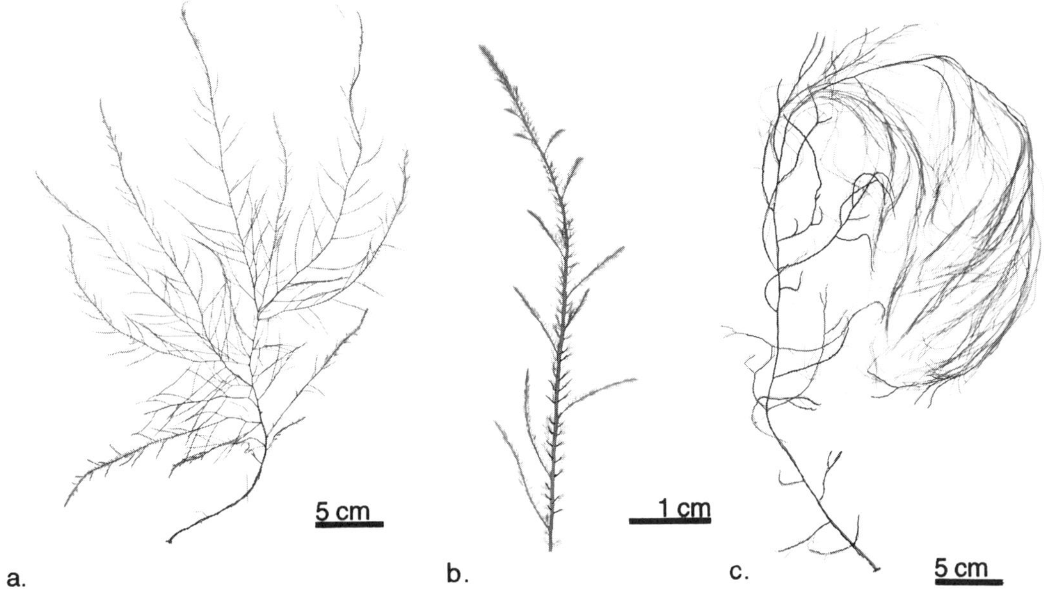

Fig. 52. ***Desmarestia aculeata***
a. young specimen with hairs (spring form); b. detail of branch tip with hairs; c. mature specimen without hairs (fall form)

Fig. 54. ***Arthrocladia villosa***
a. individual specimen
b. branch detail with whorls of tufts of hairs

Fig. 53. ***Desmarestia viridis***
a. young specimen with hairs (spring form); b. detail of branch with opposite branching and hairs; c. mature specimen without hairs (fall form))

58. **Surface smooth and slippery; thallus pale brown**; common in summer, particularly south of Cape Cod, in bays and estuaries, in sheltered areas, at or just below the low tide level, usually epiphytic on other algae, but often found floating .... *Sphaerotrichia divaricata* (Fig. 55)

58. **Rough surface produced by short tufts of hairs** (visible with a hand lens); often entangled, somewhat stiff and brittle; found in quiet, protected, shallow waters of estuaries and bays, attached to rocks and shells, or epiphytic on other algae; often found floating .... *Stilophora rhizodes* (Fig. 56)

59. **Thallus coarse, dark brown to black, abundantly branched, up to 1 m long or more; branching opposite, or alternate with flattened areas, and with pointed teeth** 60

59. **Thallus otherwise** 61

60. **Branching alternate, with flattened areas; end branches with short pointed teeth at regular intervals; thallus wiry and stiff** (see p. 66, step 56 for young form, with hairs) .... *Demarestia aculeata* (hairless, old, adult form) (Fig. 52)

60. **Branching opposite; thallus remaining relatively soft** (see p. 66, step 56 for young form, with hairs) .... *Desmarestia viridis* (hairless, old adult form) (Fig. 53)

61. **Single, central axis; branches long, whip-like, rarely rebranched, extending beyond the end of the axis**; dark brown or black; whole individual tapering to base when young, resulting in an inverted Christmas tree shape; common year round, on the open coast and in estuaries, in the interdidal and subtidal zones, down to -8 m; attached to rocks, wood and shells, or epiphytic on other algae .... *Chordaria flagelliformis* (Fig. 57)

61. **Branches conspicuously rebranched** 62

62. **One to several short main axes, with long, curving primary branches, longer than the axes;** rebranching to 3rd and 4th order, end branches curved; light brown, up to 30 cm tall; found south of Cape Cod only, in shallow, relatively protected waters, attached to rocks or shells, and epiphytic on other algae .... *Acrothrix gracilis* (Fig. 58)

62. **Single main axis, usually distinct, extending to tip of thallus; branching irregular, with branches shorter than the axis and covered with short branches**; brown, remaining thin, almost hair-like and somewhat stiff; common year round, on the open coast and in estuaries, from the low intertidal zone down to -12 m, attached to rocks, or epiphytic on other algae (often on *Chordaria flagelliformis*); (♦1 of 4) .... *Dictyosiphon foeniculaceus* (Fig. 59)

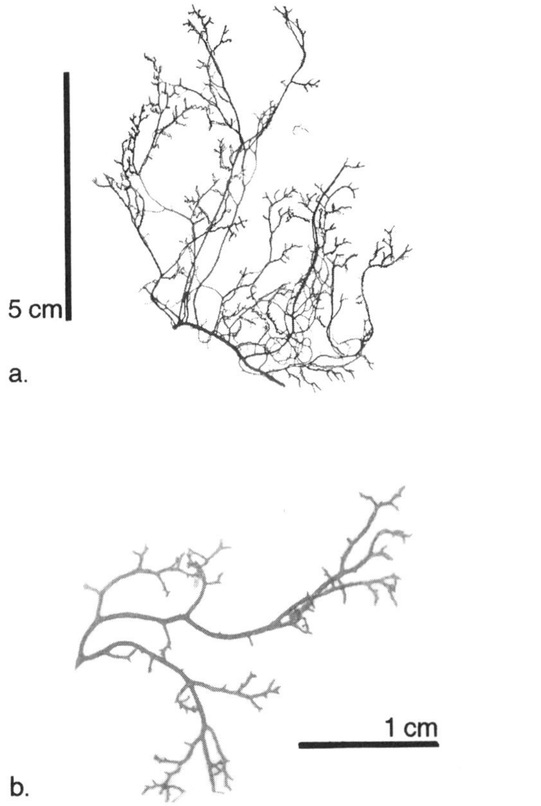

Fig. 55. ***Sphaerotrichia divaricata***
a. portion of thallus; b. detail, with branching at right angles

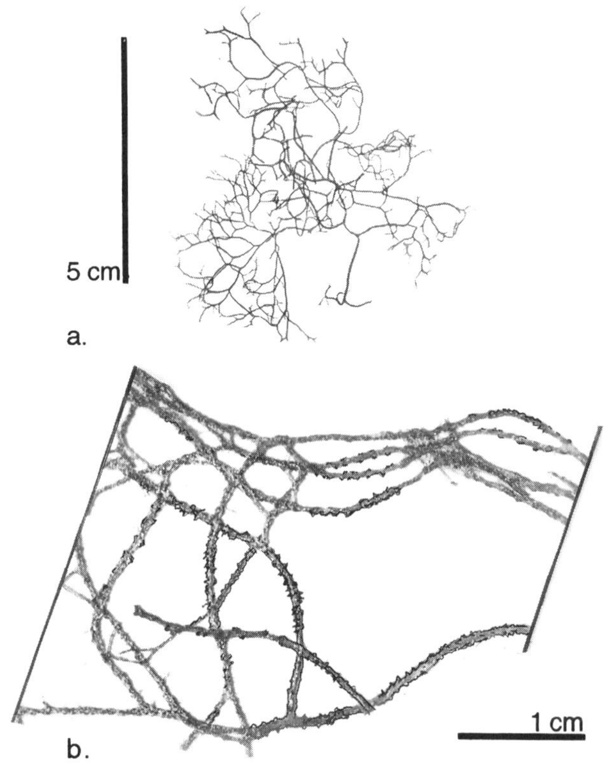

Fig. 56. ***Stilophora rhizodes***
a. whole thallus; b. detail of branches with nodules (tufts of hairs)

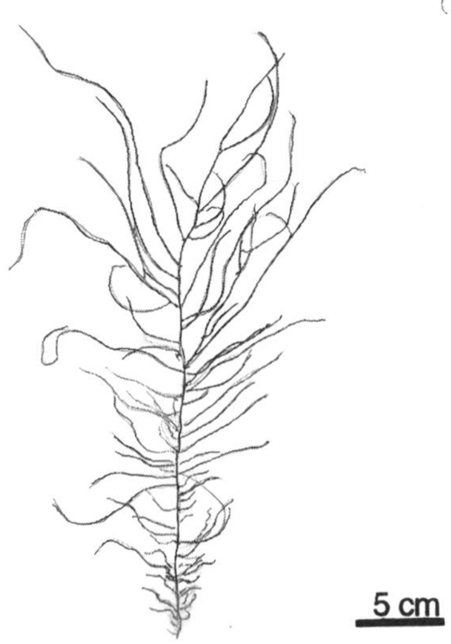

Fig. 57. ***Chordaria flagelliformis***
individual specimen

Fig. 58. ***Acrothrix gracilis***
individual specimen

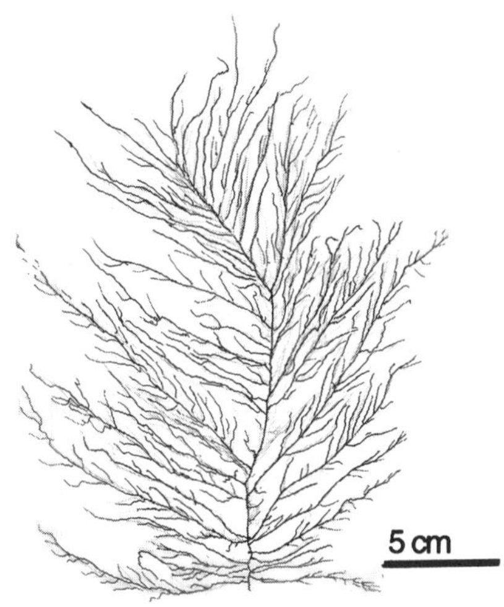

Fig. 59. ***Dictyosiphon foeniculaceus***
individual specimen

Dasya baillouviana

KEY TO THE RHODOPHYTA

Major diagnostic characters

	Page	Step
Calcareous (upright and crusts)	72-76	2-13
Fleshy crusts (non calcareous)	78-80	15-20
Very small, parasitic cushions, less than 4 mm high	82	22-24
Flat blade or tube, or with flattened blade-like parts	84-96	26-49
Wiry	98	51-55
Slippery, worm-like	100	57-58
Chains of barrel-shaped units	100	59
Hollow, soft, with constrictions at base of branches	102-104	62-65
With appendages (hairs, hooks, tendrils, pincers or proliferations on surface or at edges)	104-110	67-79
Uniseriate filaments	112-122	82-99
Polysiphonous	124-130	100-113
Thick and fleshy	132-134	114-117

1. **Thallus calcareous** (encrusted with lime, including calcium and magnesium carbonates), very hard; usually pink, but ranging from purplish red to chalky white* 2
1. **Tallus not calcareous** 14

2. **Upright, stiff tuft; beaded axes formed by a chain of calcified, articulated segments**; segments cylindrical or flattened and somewhat fan shaped; up to 5 cm tall; found year round, on the open coast and in estuaries, attached to rocks, in the low intertidal and upper subtidal zones; also found attached in tide pools, and in the drift .... *Corallina officinalis* (Fig. 1)

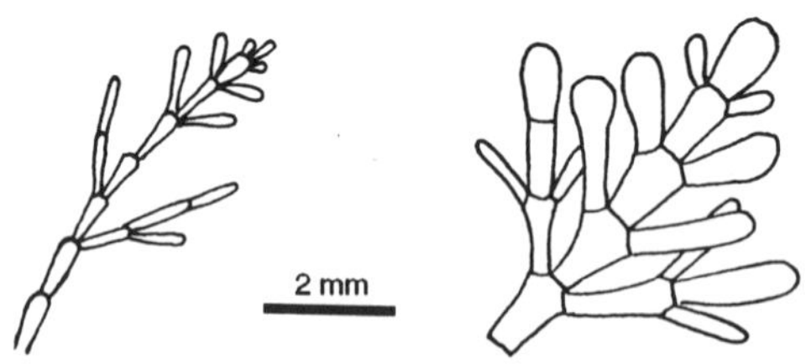

C. officinalis, segmentation types

2. **Calcareous crust or disc;** attached to rocks, shells, other algae and eelgrass; identification is often based on conceptacles characteristics; conceptacles are visible with the naked eye as spots or bumps on the crust surface, but are best seen with a hand lens; details may require a vertical section 3

3. **Stone-like crust attached to rocks, shells or other hard substrata** 4
3. **Fragile disc or crust, mostly epiphytic on other algae or eelgrass** 10

4. **Basal crust forming short, stubby, upright, and irregular, bud-like projections;** projections often branched; crusts may form large, thick masses; thallus pale to dark pink, with whitish tips; found from Cape Cod northward, on the open coast and in estuaries, in the lower intertidal and subtidal zones, down to -40 m or more; pieces of crust with projections may be found loose on the shore line, with or without substrate; (♦1 of 3?) .... *Lithothamnion glaciale* (Fig. 2)
4. **Basal crust without projections** 5

5. **Crusts smooth, but irregular, overlapping and very thick (up to 5 cm);** chunks of crusts can be easily broken off their substrata; crusts pale to purplish pink or yellowish; conceptacles (whitish dots) in clusters (sori); found from Cape Cod northward, on the open coast and in estuaries, in tide pools in the mid and low intertidal and in the sublittoral zones, down to -15 m or more, attached to rocks and shells .... *Clathromorphum circumscriptum* (Fig. 3)
5. **Crust thin (no more than 2 mm thick), difficult to remove from substrate** 6

* When cellular details or shape of conceptacles are needed, specimens may be decalcified by placing them in a weak solution of hydrochloric acid (5% HCl). This will dissolve the lime and allow sectioning.

73 (Red algae)

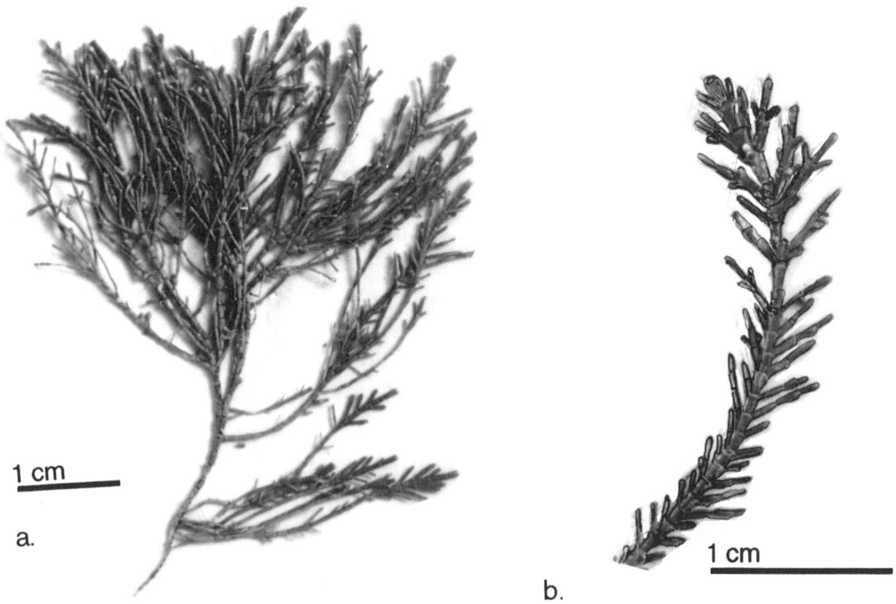

Fig. 1. ***Corallina officinalis***
a. individual specimen;
b. detail of branch

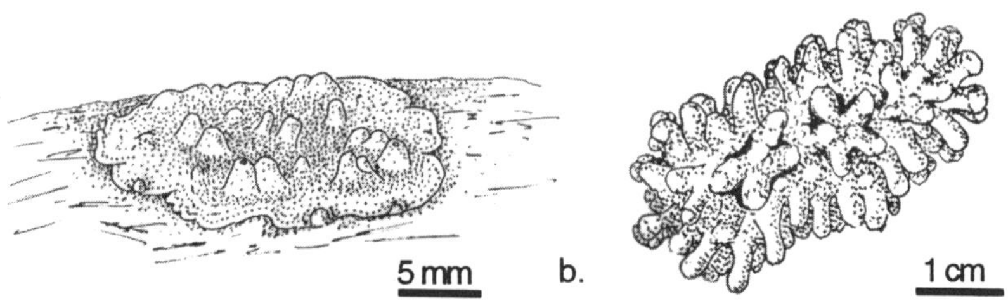

Fig. 2. ***Lithothamnion glaciale***
a. portion of young crust on rock;
b. old crusts covering mussel shell

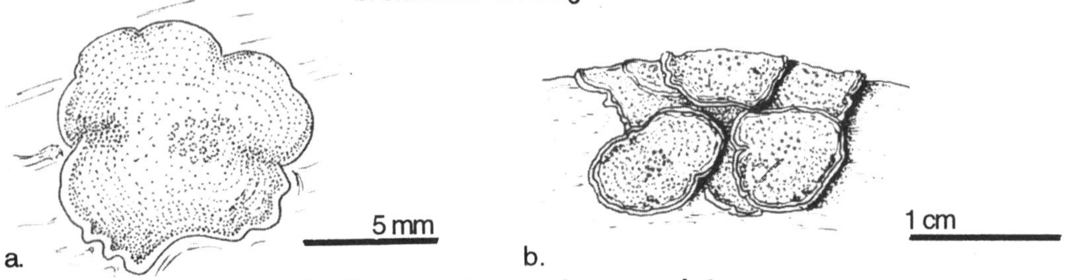

Fig. 3. ***Clathromorphum circumscriptum***
a. single crust with central sorus;
b. thick overlapping crusts on rocks

6. **Crusts with white edges and concentric rings or ridges; often overlapping;** found in both the low intertidal and in the subtidal zones 7
6. **Crusts thin, shiny and smooth; generally without ridges; rarely overlapping;** found in the subtidal zone only, mostly on shells 9

7. **Single crusts thin (less than 200 μm thick); vegetative surface smooth, becoming rough when reproductive;** thallus pink to purplish; conceptacles clustered, slightly raised, with a rounded top; common, year round, on the open coast and in estuaries, mostly in the intertidal zone, under fucoids or in tide pools, but also found in the subtidal zone, down to -3 m, attached to rocks and shells, ... *Phymatolithon lenormandii* (Fig. 4)
7. **Single crust thicker (.5-4 mm); surface uneven, with obvious ridges and white edges;** conceptacle roof sunken or depressed (Fig. 6) 8

8. **Crust with coarse ridges and conspicuous white edges; up to .75 mm thick;** thallus pink to reddish, with irregular outlines and rounded ridges; conceptacles with a raised rim and a distinct white, lens shaped cap, which is shed as a unit, at maturity; common year round, on the open coast and in estuaries, on rocks and shells, in tide pools and in the lower intertidal and subtidal zones, down to -15 m .... *Phymatolithon laevigatum* (Fig. 5)
8. **Crust with fine ridges; up to 4 mm thick;** thallus purplish pink; conceptacle opening depressed, with a thin, white, conceptacle cap, which fragments at maturity; found year round, on the open coast, in the subtidal zone only, down to -15 m or more, attached to rocks and shells .... *Phymatolithon lamii* (Fig. 6)

9. **Crust up to 2 mm thick, purplish pink; conceptacles with a depressed center or with a rim, with 40-60 pores;** found year round, north of Cape Cod, on the open coast, in the sublittoral zone, attached to rocks and shells .... *Leptophyllum foecundum* (Fig. 7)
9. **Crust up to 700 μm thick, reddish pink; large, superficial or slightly raised conceptacles (.5-1 mm wide), with 80-120 pores;** found year round, from Connecticut northward, on the open coast and in estuaries, attached to rocks and shells, in the subtidal zone only .... *Leptophyllum laeve* (Fig. 8)

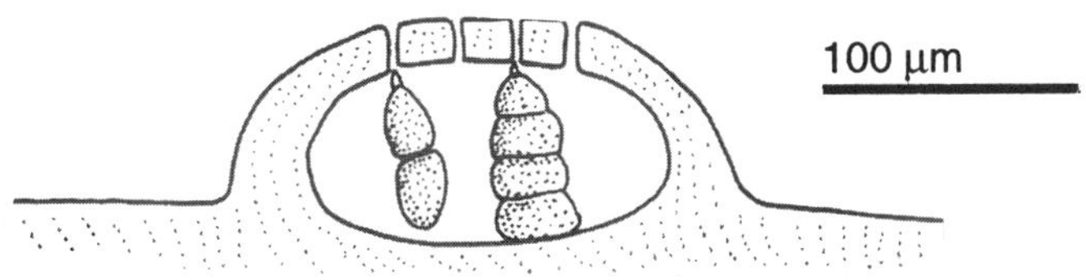

Fig. 4. ***Phymatolithon lenormandii***
vertical section of asexual conceptacle with spores

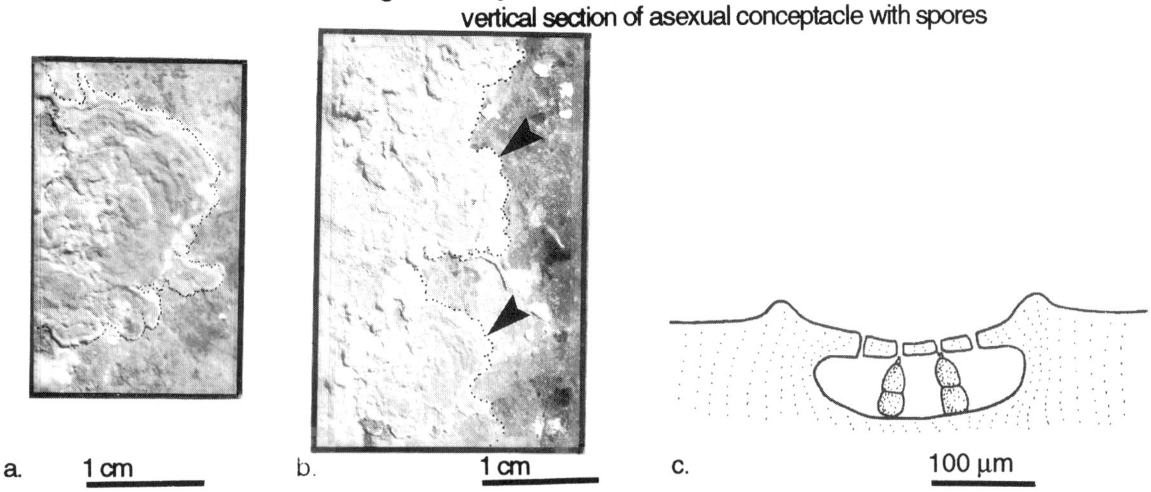

Fig. 5. ***Phymatolithon laevigatum***
a. young specimen with ridges attached to rock
b. old, bleached specimen attached to rock
c. vertical section of asexual conceptacle with spores

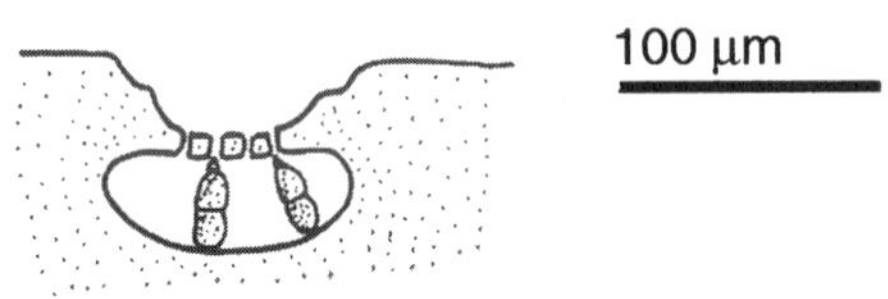

Fig. 6. ***Phymatolithon lamii***
vertical section of asexual conceptacle with spores

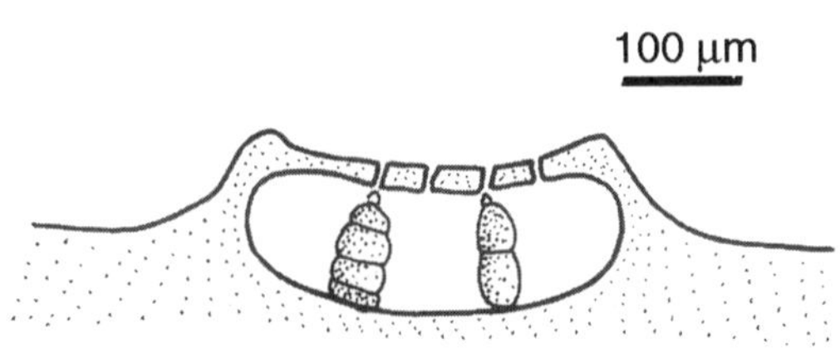

Fig. 7. ***Leptophyllum foecundum***
vertical section of asexual conceptacle with spores

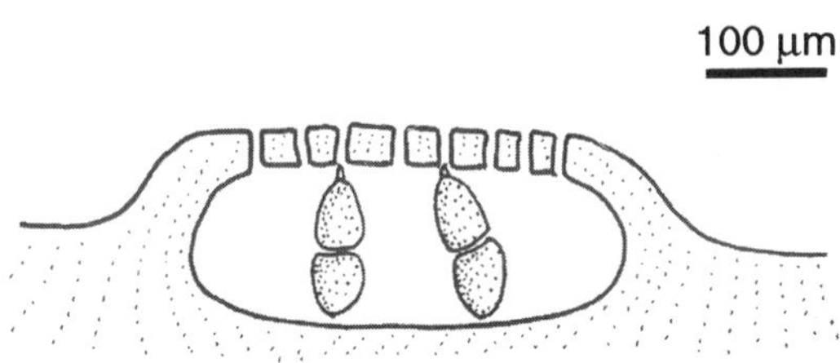

Fig. 8. ***Leptophyllum laeve***
vertical section of asexual conceptacle with spores

(Figs. 4, 5c, 6, 7 and 8, redrawn after Adey, 1973)

10. **Several pores at top of tetrasporangial conceptacles**; (drawing below); thin, small disc, 1-2 layers of cells, except at conceptacles; discs 1-3 mm in diameter; reddish to purplish pink or white; 32 to 48-celled germinating element in center of disc, with roundish cells, regularly arranged and distinct (seen only with a microscope); colonies often overlapping; growing from mid winter to fall, on the open coast and in estuaries, epiphytic, especially on red algae and on eelgrass .. .. *Melobesia membranacea* (Fig. 9)

10. **One pore only at top of tetrasporangial conceptacle** (drawing below) 11

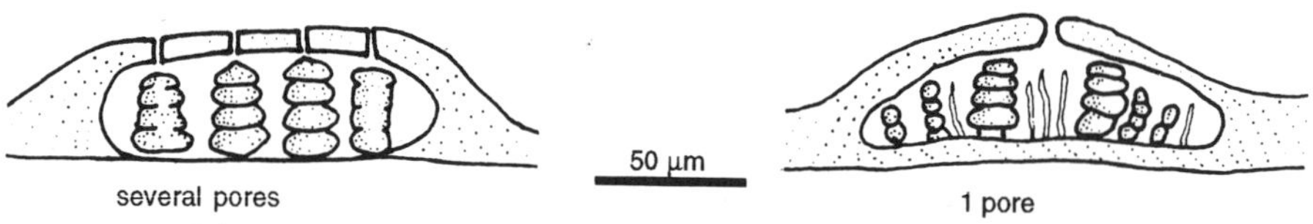

tetrasporangial conceptacles cross-sections, showing pores and tetraspores

11. **Thick, opaque crusts, 2-8 cells thick or more, heavily calcified, but remaining fragile and brittle**; roundish disc when young, becoming irregular and merging 12

11. **Thin, transparent, lightly calcified discs, easily scraped off host** 13

12. **Crusts epiphytic on other algae or eelgrass; 2-8 cells thick;** dull purplish pink,1.5-10 mm in diameter; not easily scraped off host; tetrasporangial conceptacles with one pore only; found year round, on the open coast and in estuaries, in the lower intertidal and upper subtidal zones, epiphytic on algae such as *Ascophyllum nodosum*, *Palmaria palmata*, *Chondrus crispus* and *Mastocarpus stellatus* .... *Titanoderma pustulatum* (Fig. 10)

12. **Crusts similar in appearance to the above, but epiphytic on *Corallina officinalis*; 8 cells thick or more;** purplish gray, 1-5 mm in diameter; conceptacles wart-like, with once divided sporangia; found year round, on the open coast and in estuaries .... *Titanoderma corallinae*

13. **Thin, transparent disc, 2-4 cells thick in center, one cell thick at margins; distinct, 8-celled germinating element in center of colony;** dull pink or white; 0.5-2.0 mm in diameter; tetrasporangial conceptacles with one pore only; common year round, on the open coast and in estuaries, in the subtidal zone, epiphytic, especially on eelgrass .... *Pneophyllum fragile* (Fig. 11)

13. **Thin, transparent disc, with a distinct, 4-celled germinating element in center of colony;** very similar macroscopically to the above *Pneophyllum fragile,* but the central element can be used effectively to separate the two genera; discs pinkish to white, delicate, 2-5 mm in diameter, with scattered, colorless, swollen cells; common year round, on the open coast and in estuaries, in the subtidal zone; epiphytic mostly on eelgrass, but also found on other algae .... *Hydrolithon farinosum* (Fig. 12)

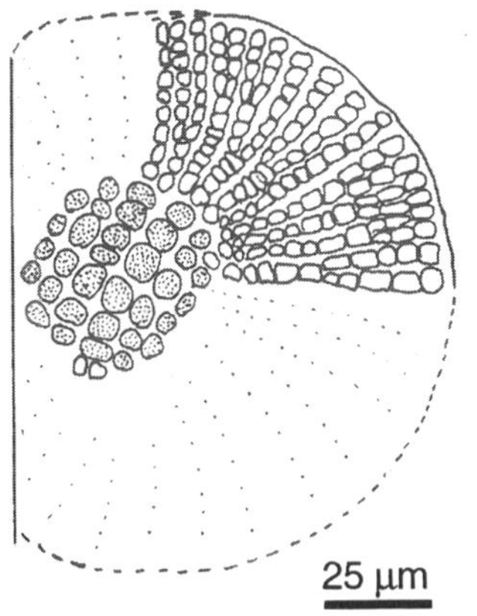

Fig. 9. ***Melobesia membranacea***
portion of colony with germinating patch

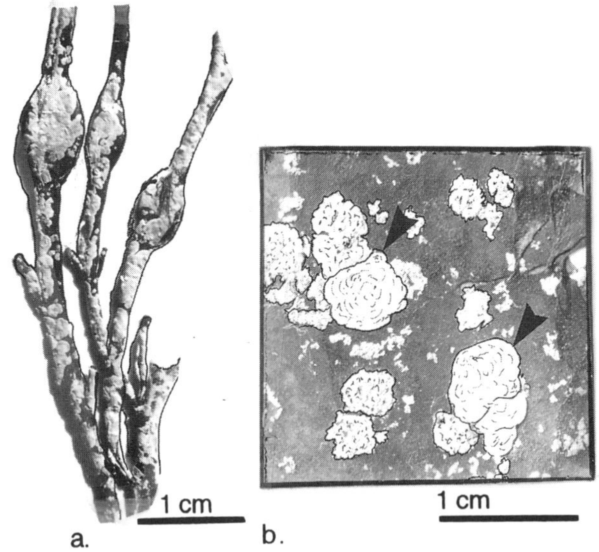

Fig. 10. ***Titanoderma pustulatum***
a. on *Ascophyllum nodosum*
b. on *Palmaria palmata*

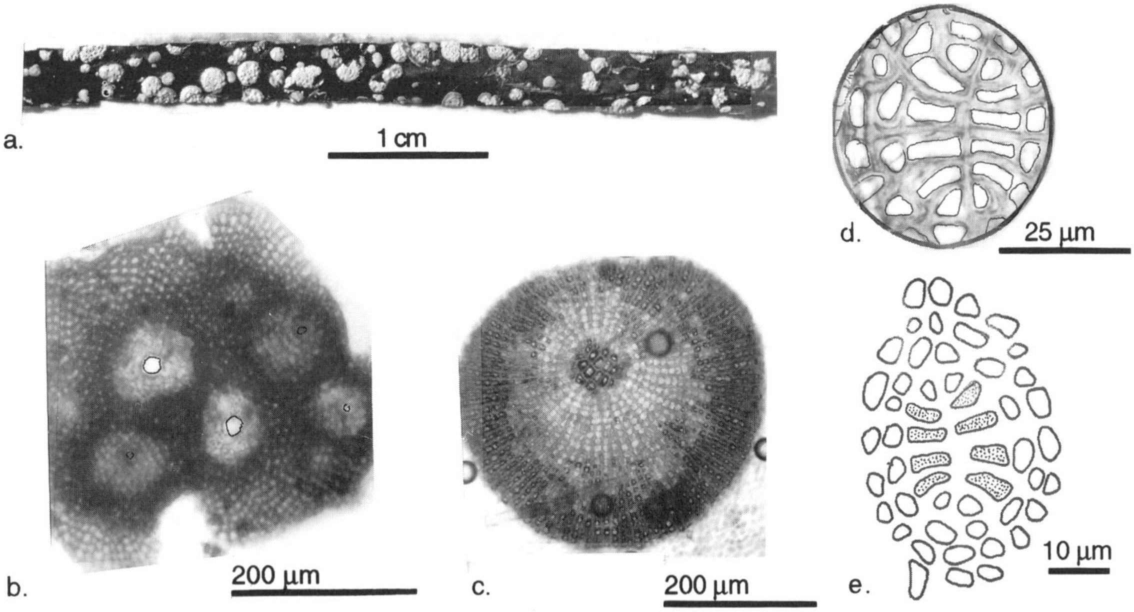

Fig. 11. ***Pneophyllum fragile***
a. colonies on eelgrass; b. colony with tetrasporangial conceptacles showing one pore; c. young colony; d. 8-celled germinating element; e. detail of germinating element

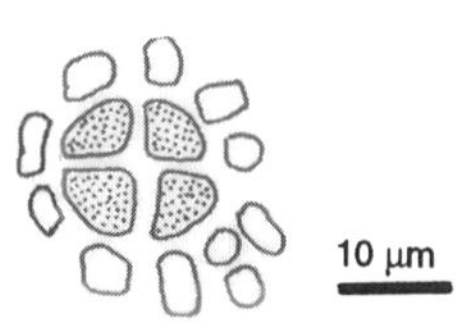

Fig. 12. ***Hydrolithon farinosum***
4-celled germinating element

14. **Non calcareous, red crust or film attached to rocks, shells, other algae or eelgrass** 15
14. **Thallus upright, or small and cushion-like, but neither crustose nor calcareous** 21

15. **Tetraspores zonate or irregular, in sunken conceptacles with clear openings (Fig. 13 d)**; openings producing a dotted surface (visible with a microscope or a good hand lens); crust or film, red, smooth and thin (up to .5 mm thick); spots roundish when young, but merging and forming irregular and extensive patches; adhering very tightly to rocks; very common year round, on the open coast and in estuaries, around the low tide level, down to -10 m, attached to rocks and shells. This is the most common of the non-calcareous red crusts found around the low tide level .... *Hildenbrandia rubra* (Fig. 13)
15. **Tetraspores not in sunken conceptacles** 16

16. **Thick crust with a velvety, slippery texture, often gelatinous; with concentric lines, but no radial striations;** thallus brownish, greenish or purplish red, .5-2 mm thick, by 5 cm or more in diameter; adhering strongly to substrate; cruciate tetrasporangia within an upright filament; found year round, on the open coast and in estuaries, mostly in the mid to low intertidal zone, but also in the upper subtidal zone, attached to rocks and shells. This crust represents the tetrasporophytic phase of *Mastocarpus stellatus* and is no longer considered a valid taxon .... *"Petrocelis cruenta"* (Fig. 14)
16. **Crust not velvety or gelatinous** 17

17. **Erect filaments forming a tight, compact layer; crust with distinct, radial striations; attached by rhizoids; slight calcification often present on the underside** 18
17. **Erect filaments loosely organized; radial striations, rhizoids and calcification absent** 19

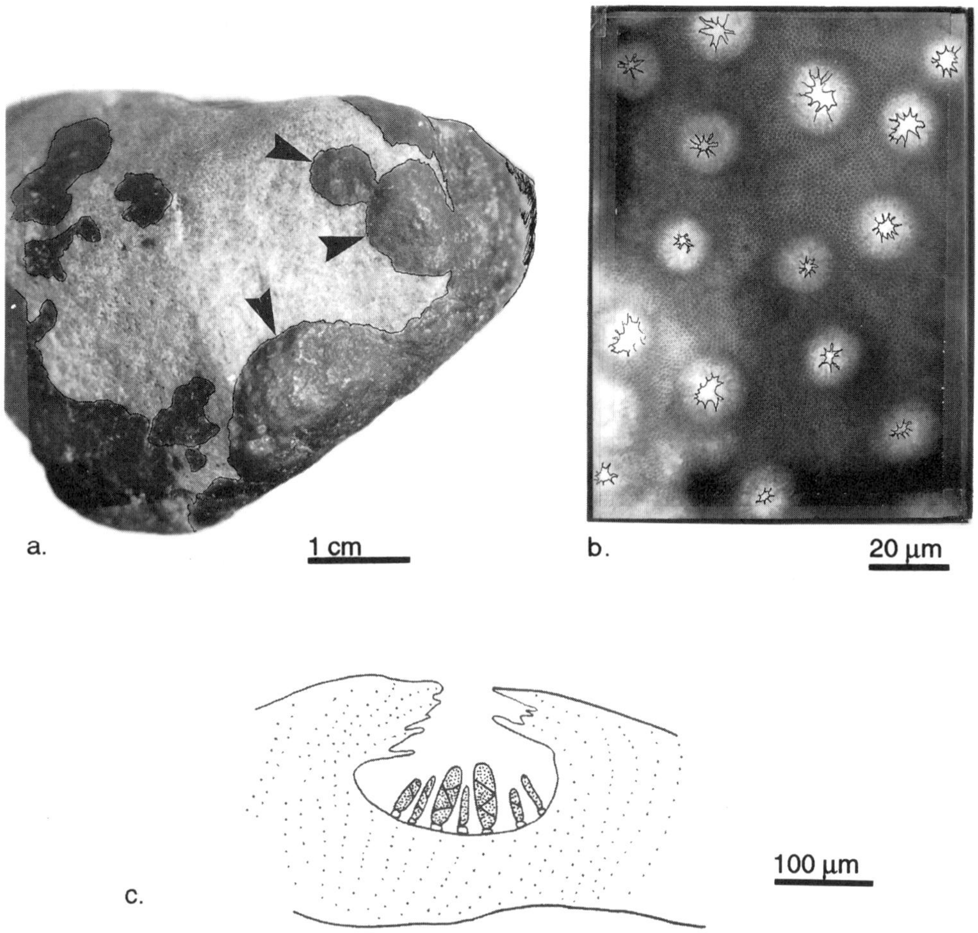

Fig. 13. ***Hildenbrandia rubra***
a. on rocks; b. openings of tetrasporangial conceptacles
c. vertical section of asexual conceptacle with tetraspores

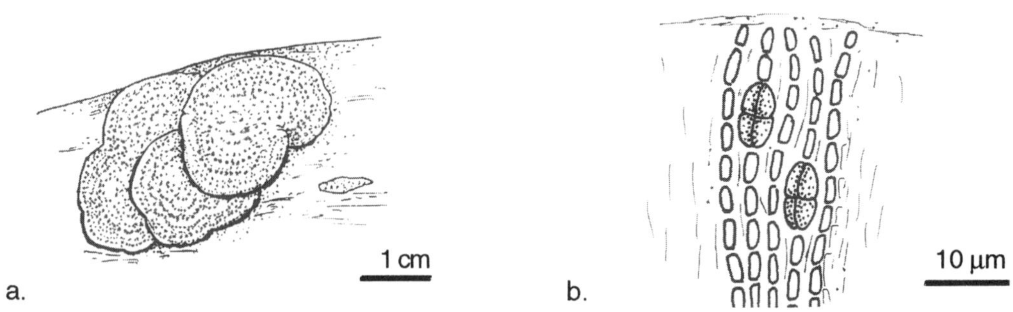

Fig. 14. "***Petrocelis cruenta***"
a. crustose sporophyte with concentric lines
b. detail of upright filaments with tetrapores

(Fig. 14a and b, redrawn after Dixon and Irvine, 1977)

18. **Crust surface smooth;** rounded patches, around 500 μm thick and up to 20 cm wide; dark brownish red to purplish red, with distinct lighter edges; found year round, from New Hampshire north, on the open coast and in estuaries, attached to rocks, or crustose corallines, mostly in the subtidal zone, down to -40 m, but also growing in low tide pools .... *Peyssonnelia rosenvingii* (Fig. 15)

18. **Crust surface wrinkled; edges often lobed;** irregularly rounded patches, around 400 μm thick and up to 5 cm wide, dark red, occasionally brownish or purplish, glossy when young; older crusts scaly; common year round, on the open coast and in estuaries, in low tide pools and in the subtidal zone, down to -16 m, attached to rocks and shells, or epiphytic on crustose corallines, and occasionally on other algae .... *Peyssonnelia dubyi* (Fig. 16)

19. **Sporangia clustered in sori (group of reproductive organs) and surrounded by curved paraphysis (sterile hairs associated with reproductive organs)** 20

19. **Curved paraphysis absent;** thin crust, up to 20 μm thick and up to 3 mm wide; cruciate tetrasporangia on stalk cells, scattered on surface of crust; found year round, on the open coast and in estuaries, in the subtidal zone, attached to shells or rocks; this crust may represent the tetrasporophytic phase of *Gloiosiphonia capillaris* and its taxonomy and position are in transition and uncertain .... "*Cruoriopsis* spp." (Fig. 17)

20. **Crust epiphytic on eelgrass** (mostly on edges of old blades); small, reddish patches, up to 1 mm in diameter, often merging; old crusts become swollen; tetrasporangia in convex sori, associated with curved or club-shaped paraphysis, 3-4 cells long; found year round, on the open coast and in estuaries, in the subtidal zone, down to -8 m; often found in the drift .... *Rhodophysema georgii* (Fig. 18)

20. **Crust attached to rocks, shells and other algae, but not on eelgrass;** deep red to dark red crusts, irregular, often merging, up to 4.5 cm wide; old crusts remaining flat; tetrasporangia in roundish sori, with paraphysis 4-6 cells long; found year round, on the open coast and in estuaries, in the subtidal zone, down to -10 m .... *Rhodophysema elegans* (Fig. 19)

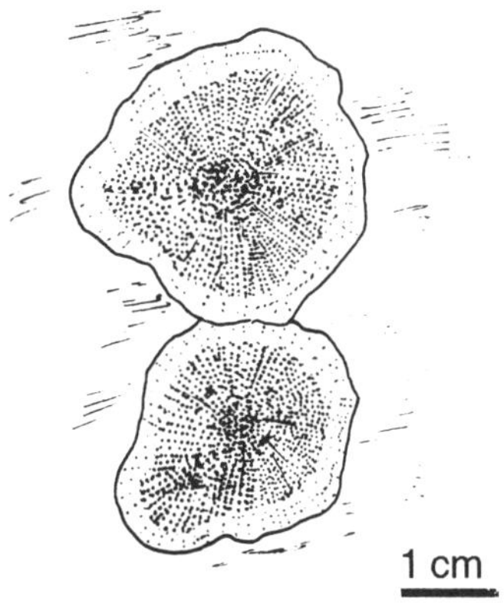

Fig. 15. ***Peyssonnelia rosenvingii***
crusts on rocks, showing radial striations and whitish edges

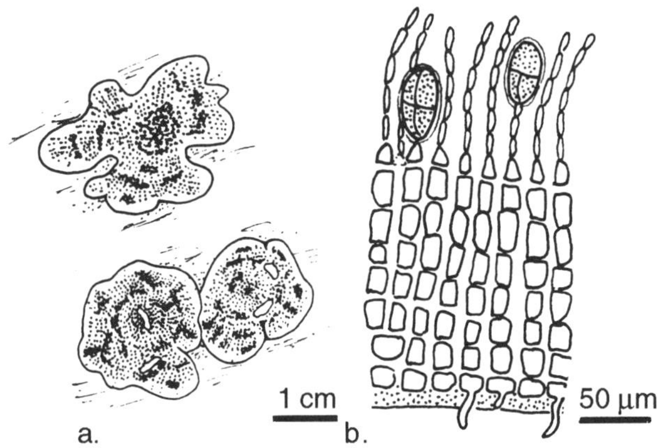

Fig. 16. ***Peyssonnelia dubyi***
a. lobed crusts on rocks, showing wrinkling and radial striations
b. detail of upright filaments with tetraspores

Fig. 17. ***Cruoriopsis* spp.**
vertical section of crust showing tetraspores attached to stalk cell at surface of crust

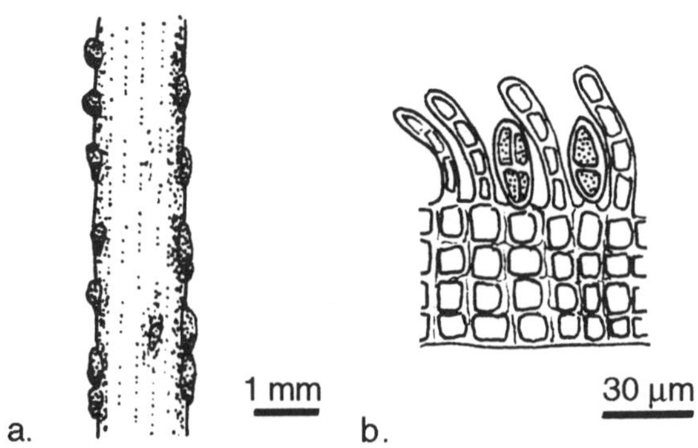

Fig. 18. ***Rhodophysema georgii***
a. crusts on eelgrass
b. vertical section of crust with tetrasporangia and paraphysis 3-4 cells long

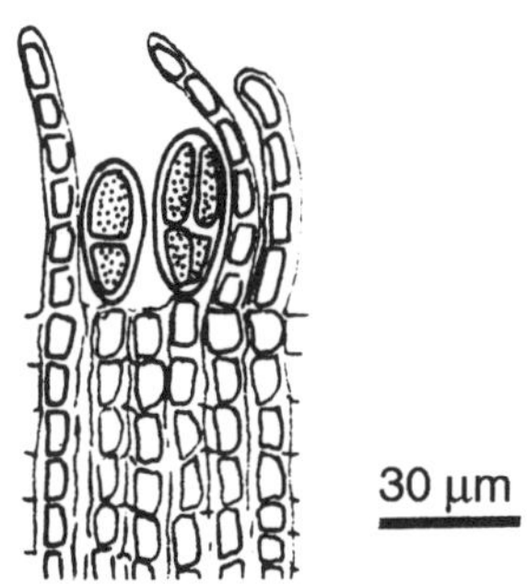

Fig. 19. ***Rhodophysema elegans***
vertical section of crust with tetrasporangia and paraphysis 4-6 cells long

(Figs. 15, 16a and b, 18a and b, 19, redrawn after Irvine, 1983; Fig 17, redrawn after Taylor, 1957)

21. **Very small, irregular, cushion-like, or wart-like thalli, no more than 4 mm high; whitish to pale brown; parasitic on specific hosts;** care should be exercised not to confuse these species with the carposporophytic generation of some Rhodophyta, visible as a rounded lump on the female gametophyte 22
21. **Thallus otherwise** 25

22. **Thallus parasitic on *Coccotylus truncatus;*** 2.5-5 mm wide, with irregular , stout branches or lobes; found year round, on the open coast, in the subtidal zone, down to -10 m .... *Ceratocolax hartzii* (Fig. 20)
22. **Thallus parasitic on other algae;** branches absent, with or without lobes 23

23. **Wart-like cushions, parasitic on *Rhodomela confervoides;*** up to 1.5 mm wide, uncommon; found year round, on the open coast, in the subtidal zone down to -20 m .... *Harveyella mirabilis* (Fig. 21)
23. **Thallus parasitic on other algae** 24

24. **Thallus parasitic on *Callophyllis cristata*;** cushions 2-4 mm high, with irregular lobes; found year round, from Cape Cod northward, on the open coast, in the subtidal zone, down to -30 m .... *Callocolax neglectus* (Fig. 22)
24. **Thallus parasitic on *Polysiphonia lanosa*;** globular, cushion-like, or somewhat lobed, up to 1 mm wide; found on the open coast, year round, but better developed from spring to fall .... *Choreocolax polysiphoniae* (Fig. 23)

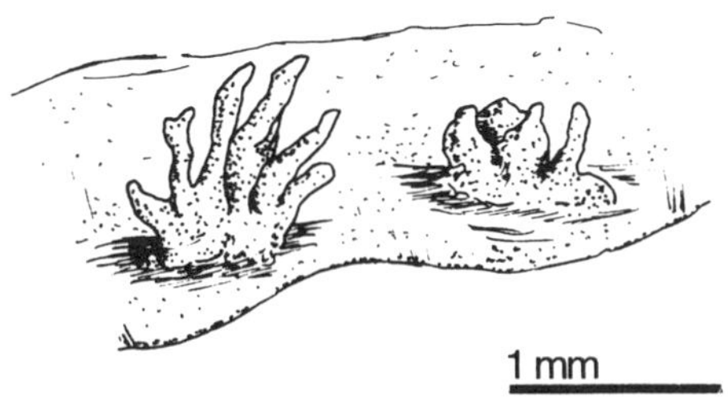

Fig. 20. ***Ceratocolax hartzii***
individual, branched specimens
on *Phyllophora* sp.

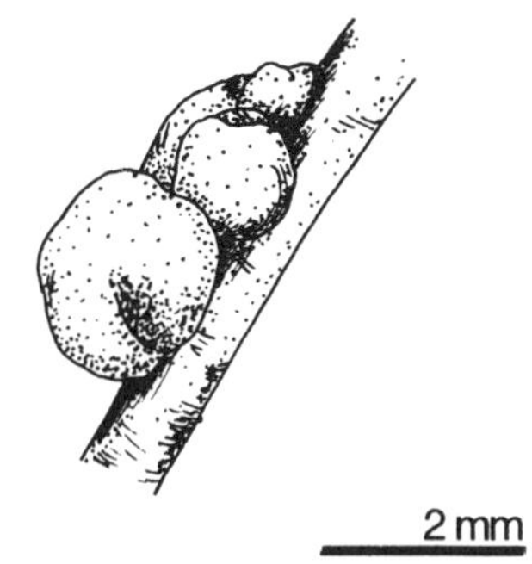

Fig. 21. ***Harveyella mirabilis***
clusters of cushions on
Rhodomela confervoides

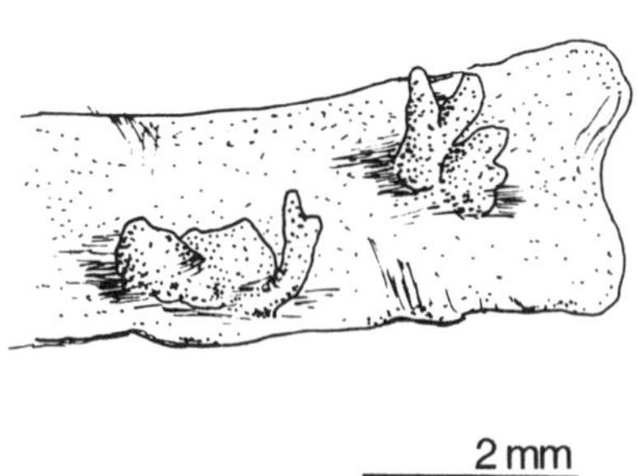

Fig. 22. ***Callocolax neglectus***
individual, lobed cushions on
Callophyllis cristata

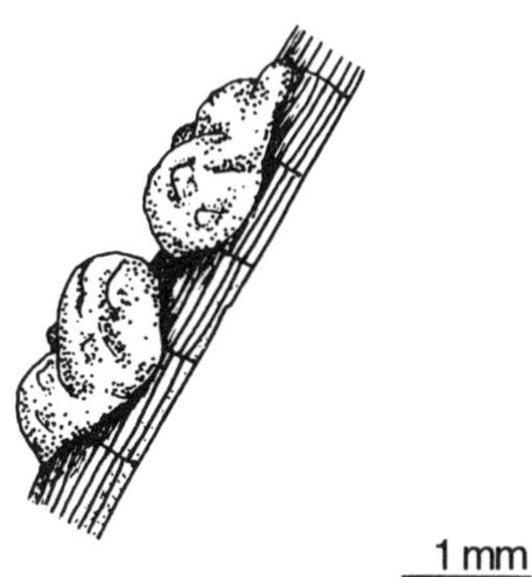

Fig. 23. ***Choreocolax polysiphoniae***
individual cushions on
Polysiphonia lanosa

25. **Thallus flattened and blade-like, or compressed in part, with flattened sections**; with or without lobes or branches 26
25. **Thallus thread-like or cylindrical for the most part (if flattened, only very narrowly so)** 50

26. **Midrib (central vein) present** 27
26. **Midrib absent** 32

27. **Thallus an undivided, delicate, pink blade, with a faint, central midrib usually extending only partway up;** almost paper-like to touch, up to 50 cm long or more; elongated or oval, with undulated edges; tips sometimes pointed; often covered with round or irregular dots (reproductive structures); found in summer only, from Cape Cod southward, in the subtidal zone, down to -12 m, mostly in estuaries, attached to rocks, wood or shells, and occasionally epiphytic on other algae; often found in the drift; turning bright orange when dying .... *Grinnellia americana* (Fig. 24)
27. **Thallus with lobes or branches** 28

28. **Leafy blade, deeply lobed and toothed at edges, with a central midrib and lateral veins;** thallus outline somewhat like that of an irregular oak leaf, up to 20 (30) cm long; pinkish red to burgundy; common year round, on the open coast and in estuaries, in the subtidal zone, down to -40 m, attached to stones and shells, or epiphytic on other algae; often found in the drift .... *Phycodrys rubens* (Fig. 25)
28. **Thallus clearly branched; branches less than 5 mm wide** 29

29. **Axes and branches flat, with regular constrictions, forming delicate chains of narrow, lance-shaped units, up to 2 mm wide, by 4-6 mm long**; midrib extending to tips of branches; lateral veins absent; thallus grayish to purplish pink, horizontally attached by rhizoids to various substrata; a warm water species, it is found in Connecticut and to the south, usually in sheltered, brackish waters, in estuaries; very tolerant of low salinities; often associated with *Bostrychia radicans* .... *Caloglossa leprieurii* (Fig. 26)
29. **Thallus without constrictions** 30

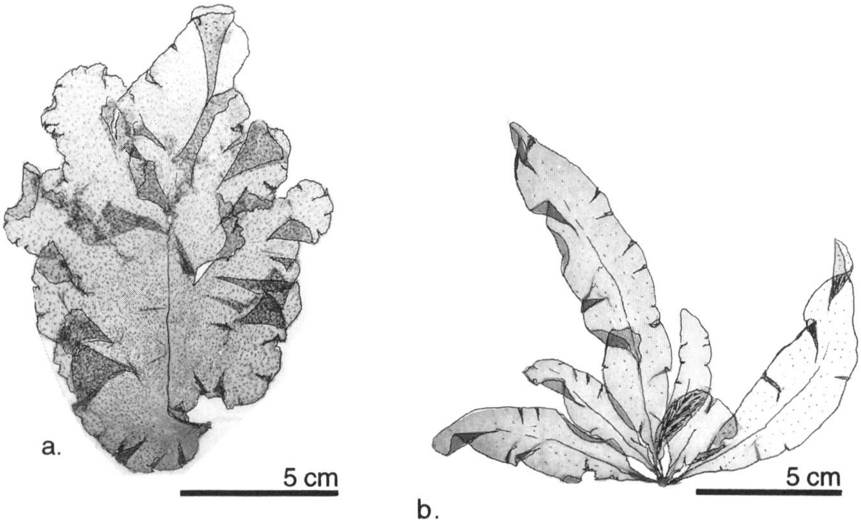

Fig. 24. ***Grinnellia americana***
a. single, oblong specimen
b. tuft of elongated specimens

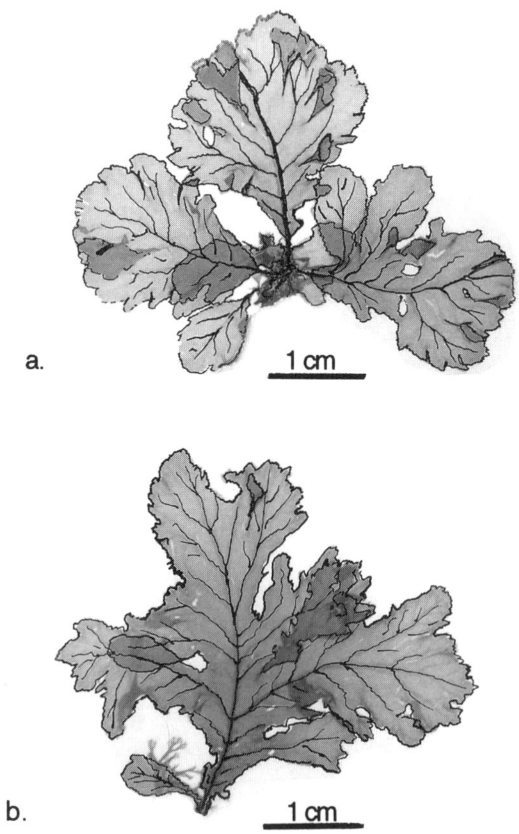

Fig. 25. ***Phycodrys rubens***
a. tuft; b. detail of deeply lobed, single specimen

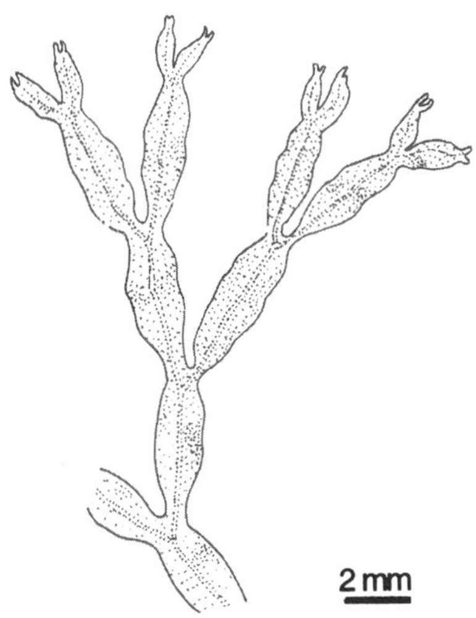

Fig. 26. ***Calloglossa leprieurii***
portion of branch showing midrib and regular constrictions

30. **Midrib, with faint lateral veins, on delicate, narrow, flattened, leafy branches (up to 4 mm wide);** thallus tufted, up to 10 cm long, rosy pink to deep red; common year round, from Cape Cod northward; found in the subtidal zone, down to -40 m, on the open coast, and occasionally in estuaries, attached to stones, or epiphytic on coarse algae; also commonly found in the drift .... *Membranoptera alata* (Fig. 27)
30. **Midrib difficult to see or only visible in transparency** 31

31. **Axes and branches visibly flattened in parts only; tips of branches macroscopically hair-like, but flattened when seen with a hand lens; no more than 2 mm wide; midrib difficult to see;** wings on either side of midrib poorly developed; thallus bushy, highly branched, up to 15 cm long; found north of Cape Cod where it is uncommon but abundant in some locations; present year round, but producing new growth in late spring and early summer, in the subtidal zone of the open coast, down to -20 m or more, epiphytic on stipes of *Laminaria* spp. and on other coarse algae .... *Pantoneura baerii* (Fig. 28)
31. **Branched midrib visible only in transparency; blades with elongated, branched proliferations, located mostly along the margins;** thallus rose to red, bushy, up to 8 (15) cm tall; flat, elongated blades, .2-2 cm wide, irregularly dichotomous; delicate, branched veins may be visible in transparency; found year round, from Cape Cod northward, mostly on the open coast, in the subtidal zone, from -5 to -40 m, attached to shells, stones or epiphytic on larger algae .... *Fimbrifolium dichotomum* (Fig. 29)

32. **Thallus a flat, undivided blade,** simple or only slightly lobed, rounded, oblong or kidney shaped 33
32. **Distinct lobes or branches present;** thallus flat or with flattened parts 37

33. **Thallus kidney shaped, more than 2 cell layers thick;** reddish brown, up to 20 cm wide, by up to 5 cm high, with numerous, pear shaped gland cells on surface; found north of Cape Cod where it is rare; growing on the open coast, in the subtidal zone, from -15 to -40 m deep, attached to rocks; collected mostly by dredging or from the drift .... *Turnerella pennyi* (Fig.30)
33. **Thallus thin, 1 or 2 cell layers thick;** gland cells absent 34

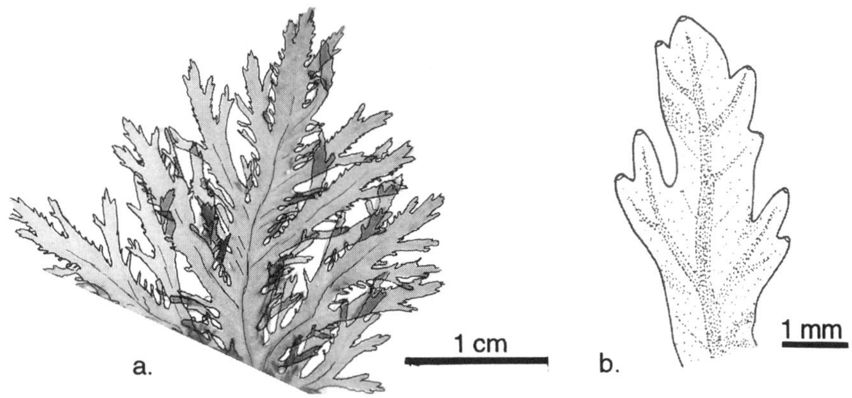

Fig. 27. ***Membranoptera alata***
a. portion of branch showing flattened branching pattern
b. detail of branch tip showing obvious midrib and wings

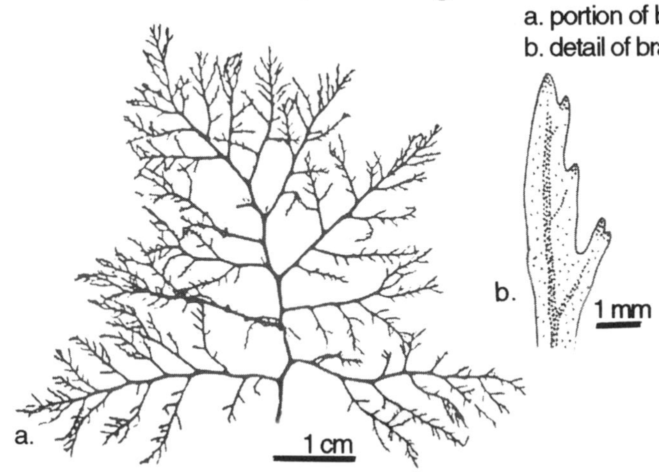

Fig. 28. ***Pantoneura baerii***
a. portion of branch showing narrowly flattened habit; b. tip of branch detail showing midrib

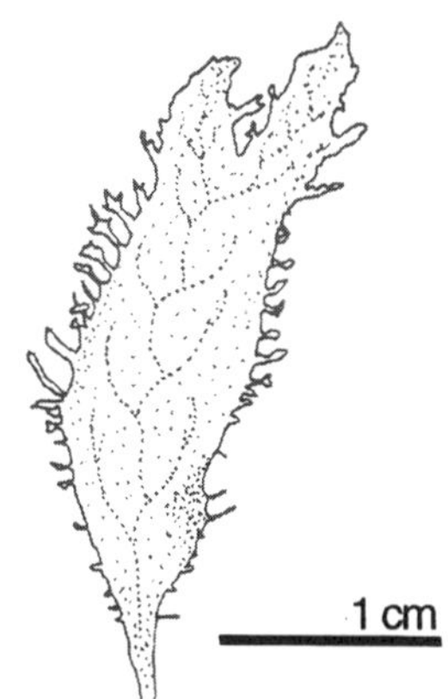

Fig. 29. ***Fimbrifolium dichotomum***
portion of branch showing proliferations and very delicate midrib visible only in transparency

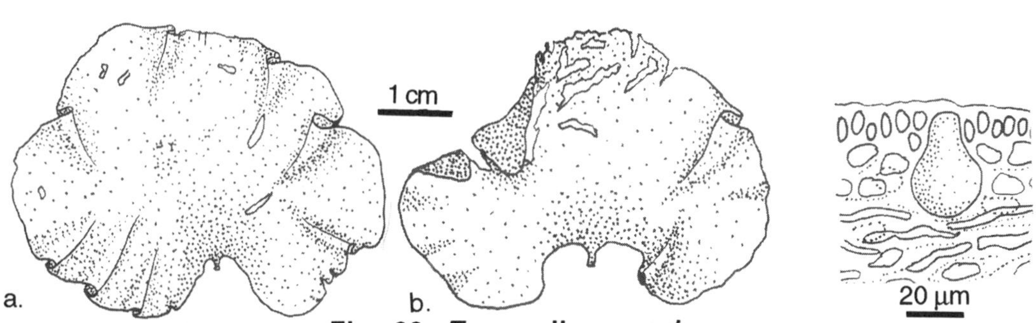

Fig. 30. ***Turnerella pennyi***
a. entire specimen; b. specimen with damaged tip
c. cross section of blade with pear-shaped gland cell

34. **Thallus 2 cell layers in cross-section of vegetative parts (drawing below);** a flat, thin blade, rounded or oblong, rose to pinkish red, or pale brownish pink; up to 30 cm (1 m) long, by up to 35 cm wide; found on the open coast and in estuaries, in the intertidal and subtidal zones, down to -10 m, from late winter to early fall, attached to rocks and invertebrates, or epiphytic on other algae; uncommon in southern New England, but much more abundant and larger, from Cape Cod northward .... *Porphyra miniata* (Fig.31)

34. **Thallus 1 cell layer in cross-section of vegetative parts (drawing below)** 35

35. **Narrow, elongated blade, less than 1.5 cm wide;** thallus reddish to yellowish brown, up to 40 cm long, by 30-50 μm thick, with undulating edges and a very small stipe; tips often shredded; found from New Hampshire northward, from late fall to late spring, on the open coast and in estuaries, in exposed areas, attached to rocks, in the mid to upper intertidal zone .... *Porphyra linearis* (Fig.32)

35. **Blade wider than 1.5 cm** 36

36. **Cells in cross-section almost square or only very slightly rectangular (diagram below); male reproductive patches whitish or yellowish and irregular;** thallus soft, slippery, sometimes rubbery; extremely variable in shape and color; may be irregularly oblong, rounded or elongated; brownish, purple or reddish color, often with yellowish tips; up to 50 cm or more; edges of blade may be scalloped or shredded; common in spring and summer, south of Cape Cod, but year round to the north; found in the intertidal zone or just below, on the open coast and in estuaries, attached to rocks, or epiphytic on other algae; this is the most common of the local species .... *Porphyra umbilicalis* (Fig. 33)

36. **Vegetative cells much higher than wide in cross-section (diagram below); blade 25-50 μm thick; male reproductive patches whitish to yellowish, elongated, and radially arranged near tip of blade;** thallus oblong, 5-20 cm long; dull pink to reddish purple or brownish; found year round but most common in spring, mostly as an epiphyte, around the mid to low intertidal zone, on the open coast and in estuaries .... *Porphyra leucosticta* (Fig. 34)

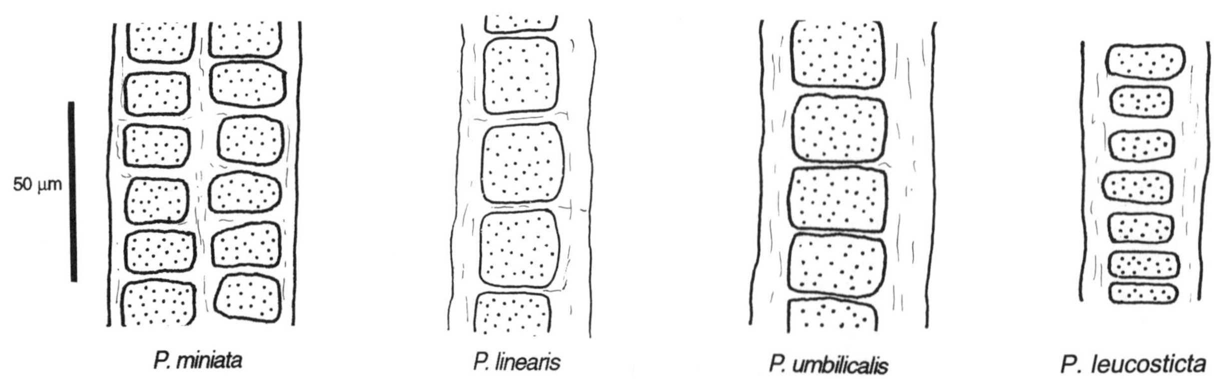

cross-sections of vegetative blades (non reproducing)

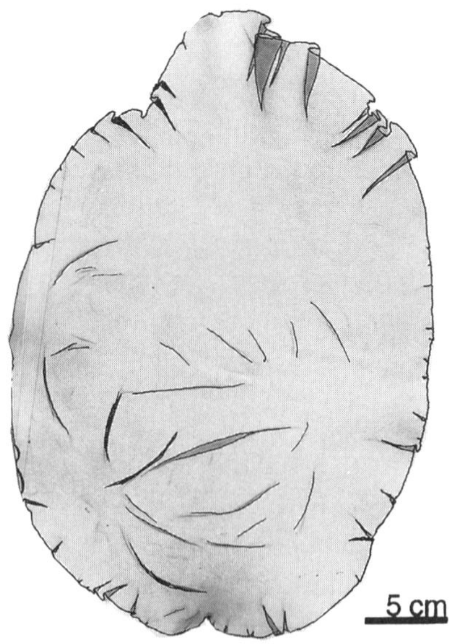

Fig. 31. ***Porphyra miniata***
typical specimen

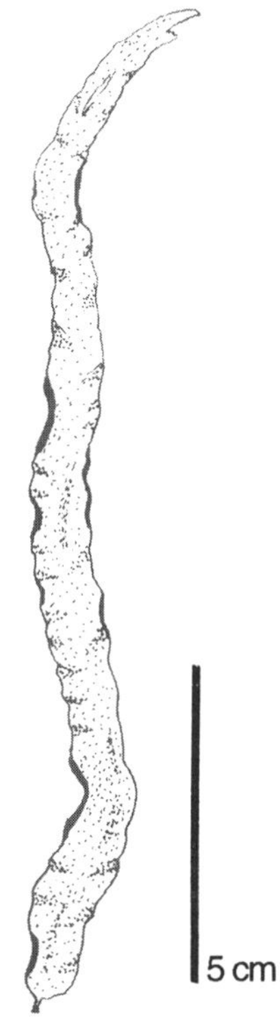

Fig. 32. ***Porphyra linearis***
typical specimen showing narrow width and undulating edges

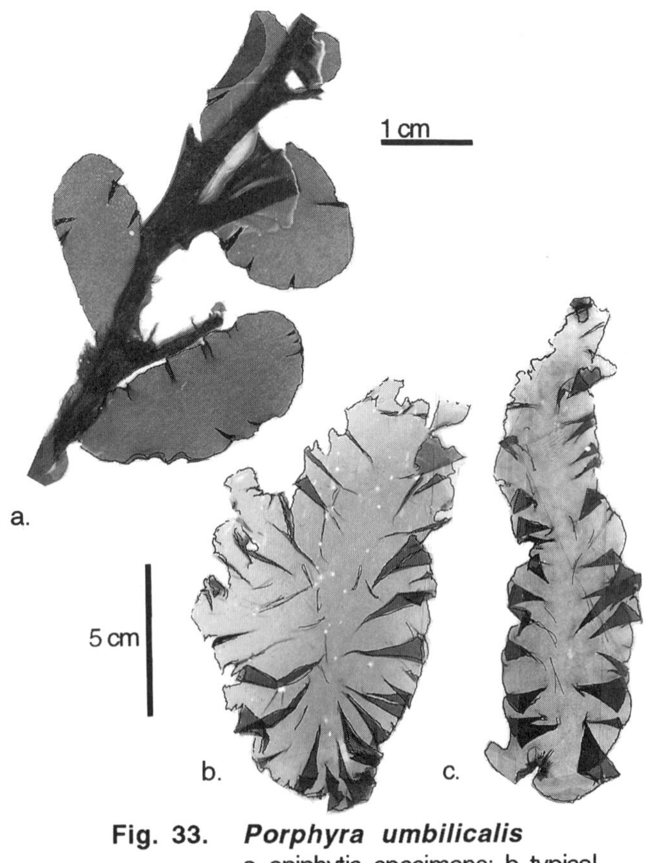

Fig. 33. ***Porphyra umbilicalis***
a. epiphytic specimens; b. typical specimen; c. elongated form

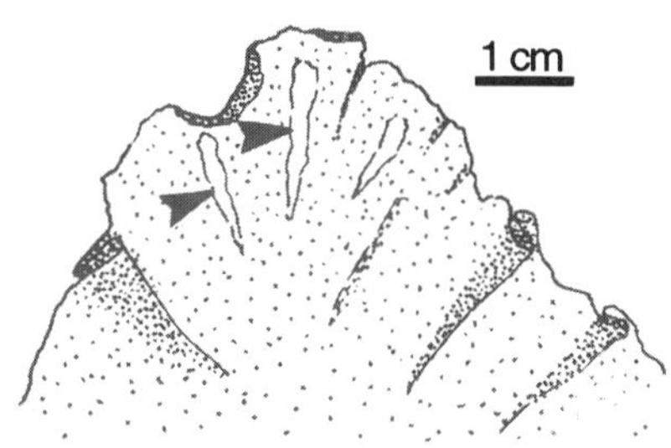

Fig. 34. ***Porphyra leucosticta***
tip of blade with reproductive patches radially arranged

37. **Large, leathery thallus, with broad lobes, 2-7 cm wide or more; stipe usually indistinct;** blade expanding gradually from a small discoid holdfast, dichotomously (Y-shaped) or palmately divided (finger-like); proliferations often present at edges; up to 50 cm long or more, pinkish red to red, occasionally greenish, yellowish or white near tips; in cross-section, blades show large, clear, medullary cells, and very small, pigmented cortical cells (see diagram below); very common year round, on the open coast and in estuaries, in the lower intertidal and subtidal zones, down to -20m, attached to rocks, or epiphytic on very coarse algae; often found in the drift .... **Palmaria palmata* (Fig. 35)

37. **Bushy thallus with a distinct stipe, or with a flat or cylindrical, distinct narrowing of the blade toward the base; branches or lobes less than 2 cm wide** 38

38. **Branches flattened, feather-like, highly dissected**; ultimate branchlets in pairs, with one of each pair reduced 39

38. **Branches flattened but not feather-like** 40

39. **Branches somewhat stiff; covered by cortication (except 1 or 2 terminal cells); tip of branch triangular when seen with a hand lens;** branching in one plane, feather-like, bilaterally opposite, with one of each pair smaller than the other; branches may have teeth or projections; thallus a bushy tuft, up to 15 cm long, somewhat stiff, rosy red to brownish red; found year round; common from Cape Cod northward; rare to the south; found in the subtidal zone, -10 to -40 m deep, on the open coast and in estuaries, often attached to the base of kelps or to rocks; frequently found in the drift .... *Ptilota serrata* (Fig. 36)

39. **Branches soft and delicate, with coarse and cylindrical main axes; tips of branches uniseriate;** thallus bushy, polysiphonous (p. 112, step 81), ranging from rosy red to brown; branching in one plane only, alternately bilateral, but opposite at tips, with one of each pair reduced; found year round, on the open coast; in the subtidal zone, in deep waters only south of Cape Cod, as well as in the low intertidal zone to the north; attached to rocks and epiphytic on other algae; common in the drift .... *Plumaria plumosa* (Fig. 37)

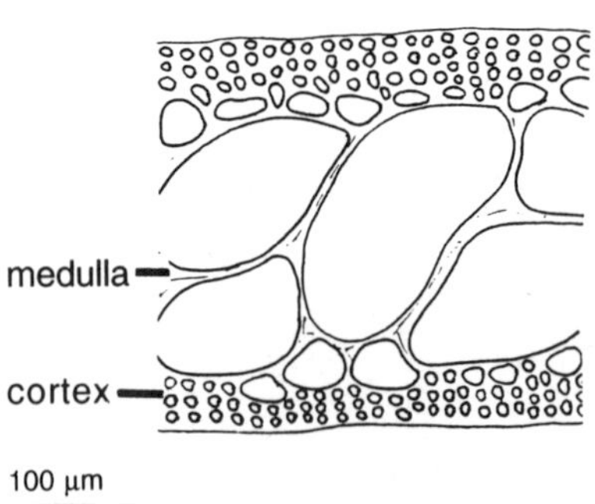

Palmaria palmata
cross-section

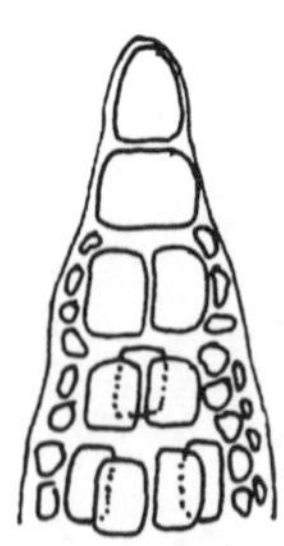

Ptilota serrata
triangular tip of branch

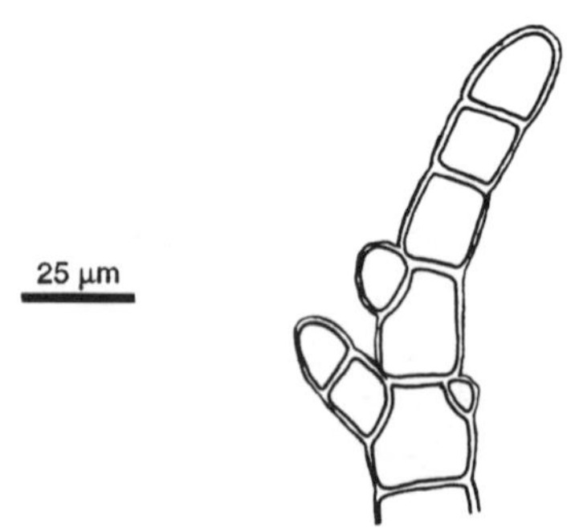

Plumaria plumosa
uniseriate tip of branch

* Since July 1996, *Grateloupia doryphora*, an invasive red alga has been recorded in Narragansett Bay, R.I. It resembles *P. palmata* and would key-out to couplet 37. It is described in the Addendum, p. 145.

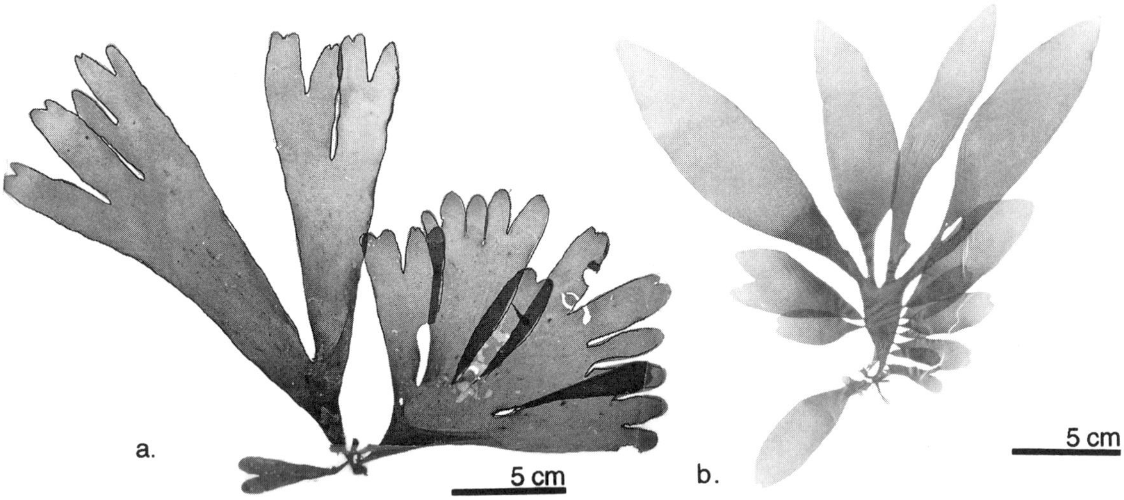

Fig. 35. ***Palmaria palmata***
a. specimen showing dichotomous lobes
b. specimen with marginal proliferations

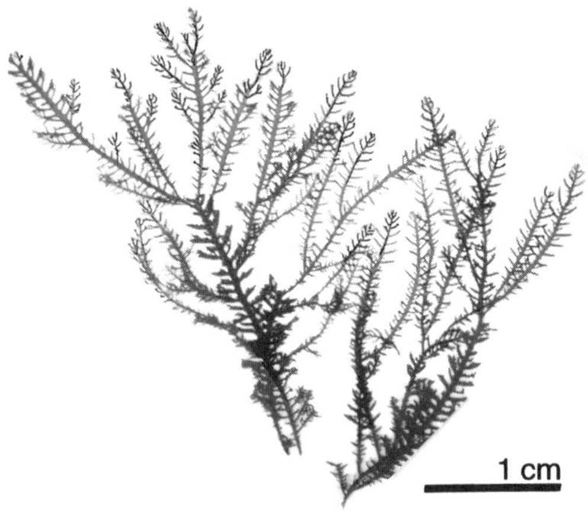

Fig. 36. ***Ptilota serrata***
portion of branches

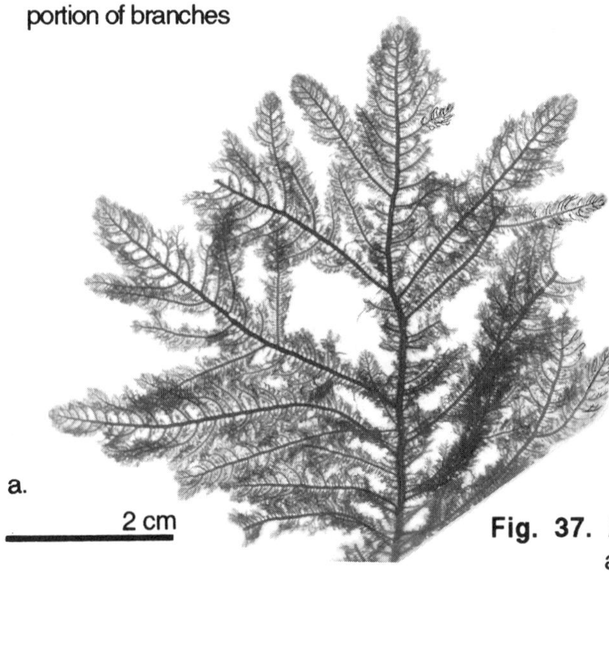

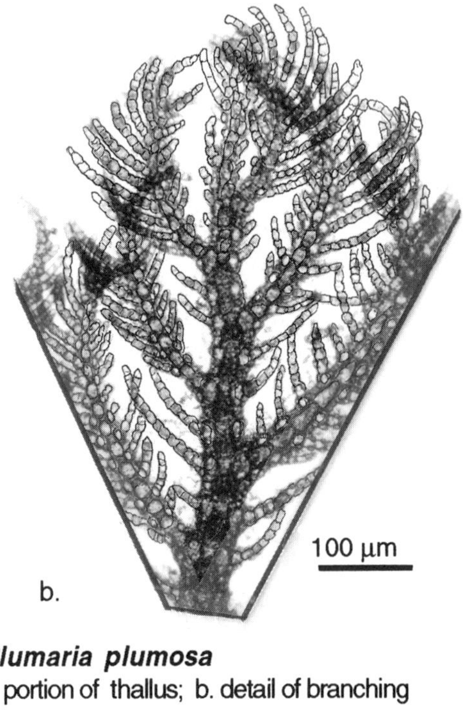

Fig. 37. ***Plumaria plumosa***
a. portion of thallus; b. detail of branching

40. **Branches and/or axes hollow throughout, or in part** 41
40. **Axes and branches solid, or not obviously hollow** 42

41. **Soft, long branches, hollow, often twisted, flattened or tubular in part;** thallus sparingly rebranched, branches tapering to base; generally 10-20 cm long, but up to 50 cm or more; dull orangish to purplish, or brownish red, becoming yellowish or greenish; very common from late winter to early summer, with some individuals persisting through the year, north of Cape Cod; found in shallow waters, on stones and shells, around the low intertidal and upper subtidal zones, down to -5 m, and in tide pools throughout the intertidal zone; often growing in clumps .... *Dumontia contorta* (Fig. 38)
41. **Stiff, firm, fleshy axes, hollow and somewhat flattened in parts;** up to 40 cm long, by up to 1.5 cm wide; extremely variable in form, ranging from unbranched, to heavily branched axes; branches may be long and simple, or consist of numerous, short proliferations; tapering to base; dark red to purplish; found year round, from Cape Cod northward, in the low intertidal and in the subtidal zones, down to -10 m, attached to rocks, on the open coast, in very exposed areas only .... *Devaleraea ramentacea* (Fig.39)

42. **Base of branches abruptly constricted; small, delicate thallus;** main axes flat, with opposite, elongated, flat branches in two rows; only occasionally rebranched; soft, bright rose, up to 3 (5) cm tall; rare species found year round, from Rhode Island to Maine, on the open coast and in estuaries, in the subtidal zone, down to -12 m, attached to rocks, stones and wood, or in the drift; also growing in low, deep tide pools (♦1 of 3) .... *Lomentaria orcadensis* (Fig. 40)
42. **Larger thallus, with base of branches not abruptly constricted** 43

43. **Flat branches regularly decreasing in size at each division;** less than 3 mm wide, repeatedly branched; thallus tufted, generally less than 8 cm high, but up to 15 cm high, pink to rosy red; found year round, from Cape Cod northward, in the subtidal zone, down to -40 m, on the open coast and in estuaries, attached to rocks and invertebrates, or epiphytic on coarser algae; often found in the drift .... *Callophyllis cristata* (Fig. 41)
43. **Branch size, either remaining the same, or gradually widening, or irregular, but not decreasing regularly** 44

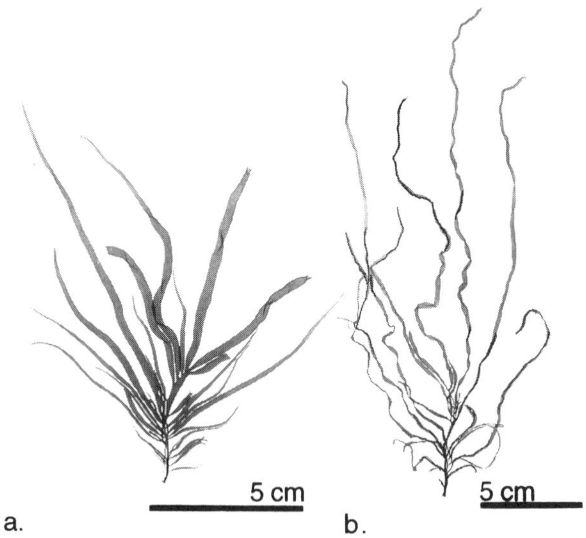

Fig. 38. ***Dumontia contorta*,**
a. young specimen, collected in early spring
b. late spring, mature specimen with twisted branches

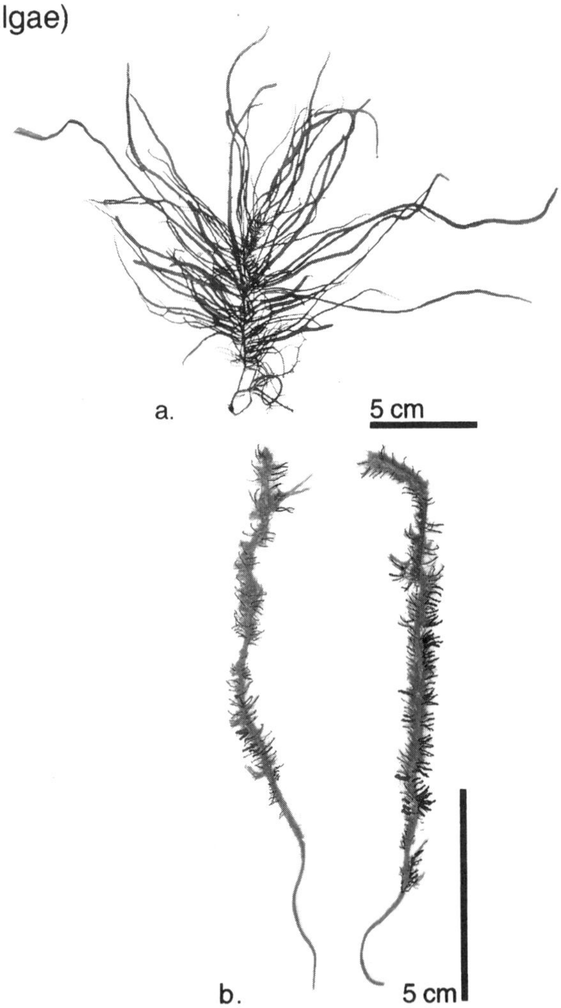

Fig.39. ***Devaleraea ramentacea***
a. branched specimen; b. single axes with proliferations

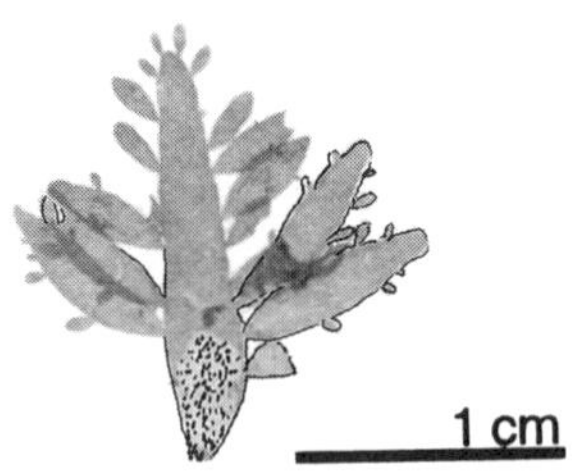

Fig. 40. ***Lomentaria orcadensis***
single specimen with constricted base of branches

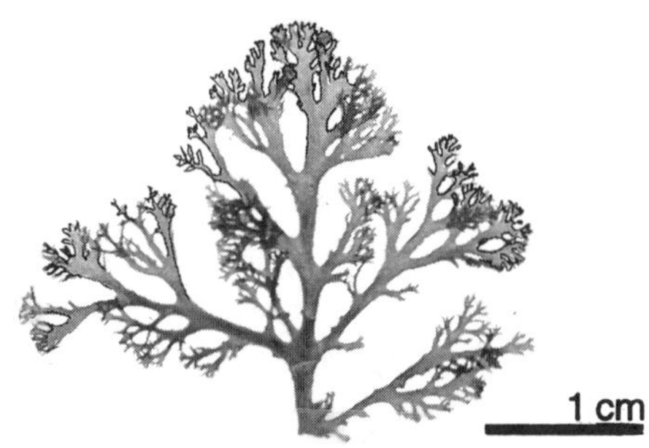

Fig. 41. ***Callophyllis cristata***
portion of branch

44. **Main axes and branches wiry, cylindrical and thin, flaring out to flat, blade-like surfaces, only in upper parts**; often encrusted with bryozoans 45

44. **Stipe flattened or cylindrical, gradually widening from base; branches or lobes in one plane** 46

45. **Blades abruptly and sharply expanding only near tip; axes long and wiry; female reproductive structures located on short stalks at base of blade** (see below); thallus up to 50 cm, rosy red when young, turning dark red to black; found year round, on the open coast and in estuaries, in the subtidal zone, down to -20 m, occasionally in low, deep tide pools, attached to rocks and shells, or epiphytic on coarse algae; common in the drift .... *Phyllophora pseudoceranoides* (Fig. 42)

45. **Branches expanding gradually to blade; female reproductive structures located on short stalks, at tips of blades** (see below); thallus up to 10 (30) cm long; rosy red when young, turning dark purplish red to black; found year round on the open coast and in estuaries, in the subtidal zone, down to -40 m, attached to rocks, shells and epiphytic on coarse algae; common in the drift .... *Coccotylus truncatus* (Fig. 43)

P. pseudoceranoides, tip of blade

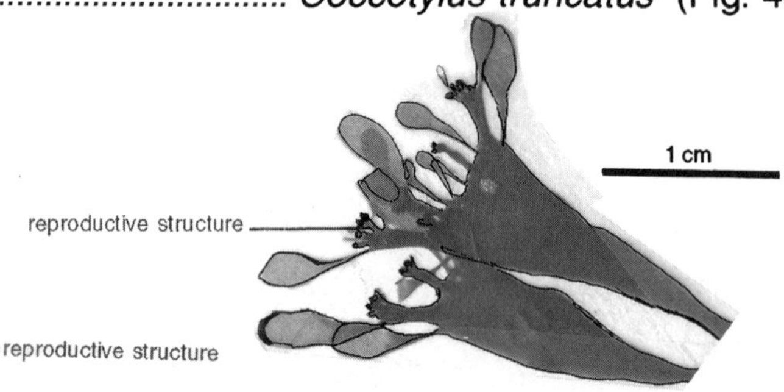

C. truncatus, tip of blade

46. **Axes and branches very irregular and often flattened, especially at forkings; cylindrical in part;** branching in a flat plane, with short, thin, pointed spur-like projections; firm, fleshy, up to 40 cm long; often with scattered prominent dark red reproductive lumps on surface; brownish red to dull purple, yellowish green or straw color, occasionally red only at base; very common and variable; found year round, in the subtidal zone, in warm, shallow, quiet waters, mostly in estuaries; often detached (see p. 132, step 116, Fig. 101, for cylindrical form, previously described as a separate species) .... *Gracilaria tikvahiae* (Fig. 44)

46. **Branches or lobes regularly or irregularly dichotomous (Y-shaped)** 47

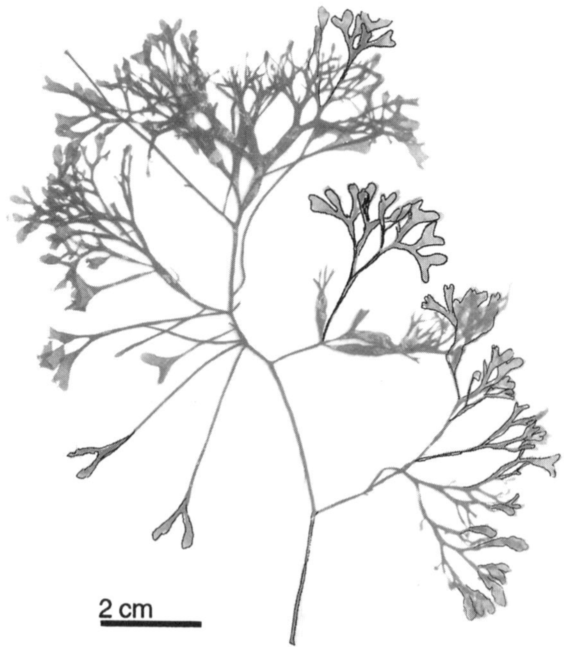

Fig. 42. ***Phyllophora pseudoceranoides***
portion of specimen

Fig. 43. ***Coccotylus truncatus***
portion of specimen

Fig. 44. ***Gracilaria tikvahiae***
portion of flattened branch with spurs

47. **Surfaces and/or margins with irregular proliferations (from small bumps to short, branched growths)** 48
47. **Irregular proliferations absent** 49

48. **Small, irregular, rounded or elongated proliferations or bumps, located mostly on the surface,** 1 mm wide by up to 2 mm long; thallus firm, thick and fleshy, upright, 3-10 cm tall; stalk-like base often with rolled edges, flattening out to mostly dichotomous but sometimes irregular branching; dark purplish red to almost black; found year round, mostly on the open coast, in exposed areas, attached to rocks around the low tide level, and in mid to lower intertidal tide pools .... *Mastocarpus stellatus* (Fig. 45)
48. **Elongated, branched proliferations, located mostly along the margins of flat, elongated blades, .2-2 cm wide** (see p. 86, step 31 for complete description) .... *Fimbrifolium dichotomum* (Fig. 46)

M. stellatus
tip of blade with proliferations (papillae)

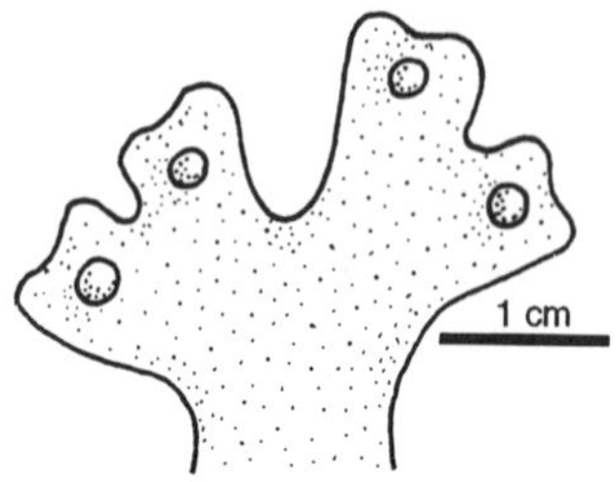

C. crispus,
tip of blade with reproductive structures (nemathecia)

49. **Flat stipe, flaring out to flat, smooth, dichotomous branches (2-15 mm wide) with blunt or rounded tips; medulla somewhat loose and filamentous** (drawing below); thallus upright, bushy and leathery, 5-10 cm tall, occasionally longer; with regular, 3-5 mm, round mounds near tips, present only when reproductive (drawing above); shape and color extremely variable; from dark purple red in deep waters, to bleached yellowish green in exposed areas; one of the most common species, found year round, on the open coast and in estuaries, in the low intertidal and subtidal zones, down to -10 m, attached to rocks and shells; often growing in clumps or large populations .... *Chondrus crispus* (Fig. 47)
49. **Stipe cylindrical, gradually flattened out into blade-like branches; medulla of large, roundish or squarish cells, relatively tightly packed;** (drawing below); branching dichotomous, up to 3 mm wide, with the two edges of a blade parallel to each other; thallus pink when young, but brownish red, with pink tips, at maturity, up to 5 (10) cm long; found year round, in southern Massachusetts and to the north, on the open coast and in estuaries, attached to rocks, in low, deep tide pools, or in the subtidal zone, down to -15 m; often floating in the drift .... *Gymnogongrus crenulatus* (Fig. 48)

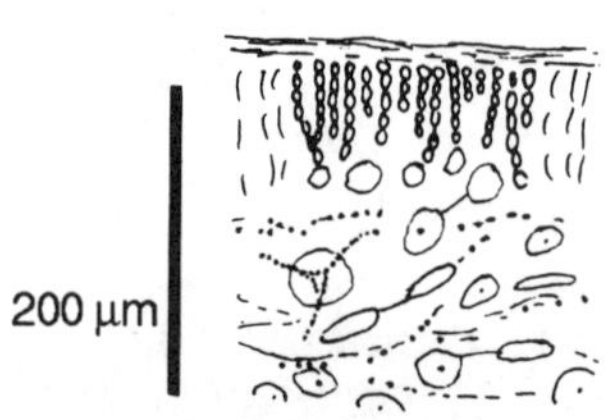

C. crispus, portion of cross section

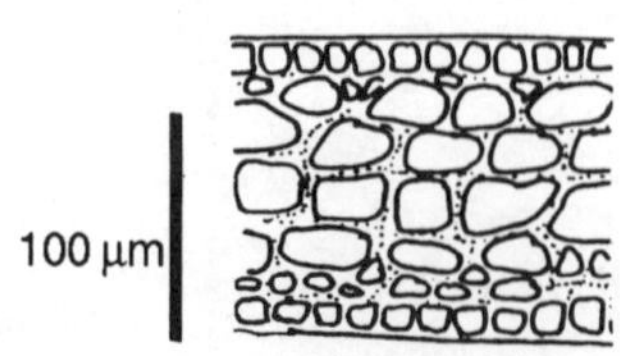

G. crenulatus, cross section

Fig. 45. ***Mastocarpus stellatus***
branch with surface proliferations

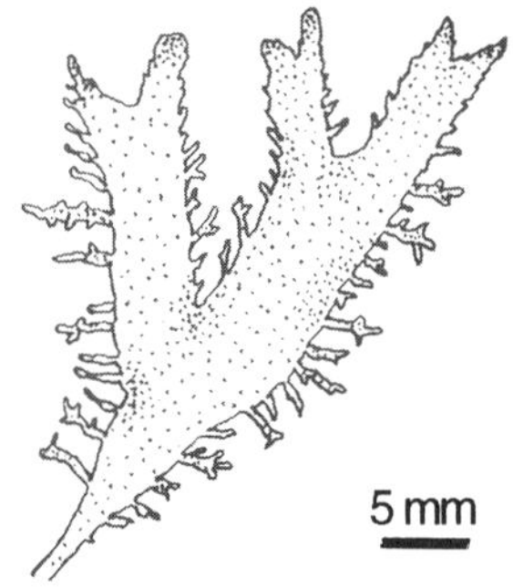

Fig. 46. ***Fimbrifolium dichotomum***
branch with marginal proliferations

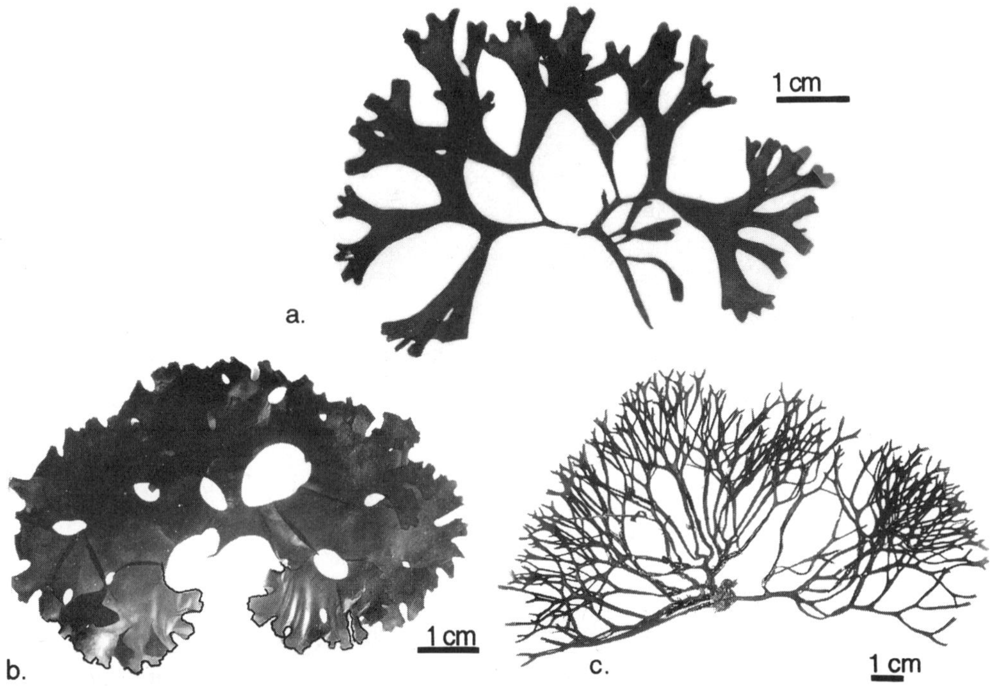

Fig. 47. ***Chondrus crispus***
a. typical form; b. broad form from a shallow estuary;
c. narrow form from a deep subtidal zone, on the open coast

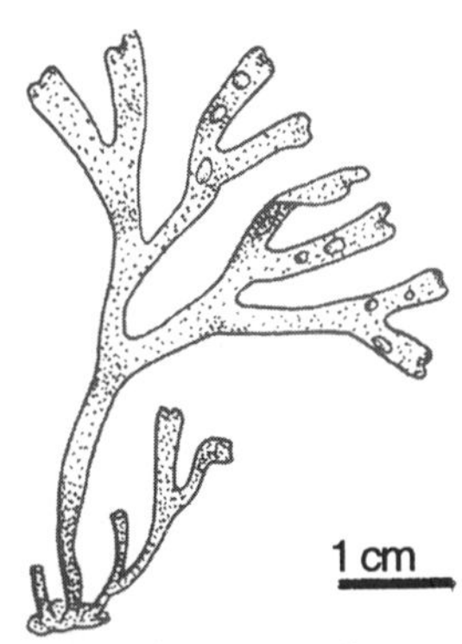

Fig. 48. ***Gymnogongrus crenulatus***
whole specimen with
reprodutive structures

50. **Thin, wiry, bushy thallus, almost black; coarse or delicate** 51
50. **Thallus not wiry; may be coarse or delicate** 56

51. **Short, stiff tufts, epiphytic on *Ascophyllum nodosum*;** branches dense, dichotomous (Y-shaped) and thin, less than 0.5 mm wide; spherical tuft, up 2-5 cm tall; with a typical *Polysiphonia* construction: a uniseriate filament covered with 20-24 pericentral cells (see p. 112, step 81); dark red or black; found year round, in the lower and mid intertidal zone, on the open coast and in estuaries; (♦1 of 10) .... *Polysiphonia lanosa* (Fig. 49)
51. **Thallus attached to other substrata, or cast ashore in the drift** 52

52. **Tufts of fine branches present during spring and early summer** (see p. 124, step 101, Fig. 90) .... *Rhodomela confervoides*
52. **Ultimate branches not tufted** 53

53. **Short tufts, no more than 5 cm high,** attached to rocks around the low tide level 54
53. **Tufts longer, up to 20 cm high;** in tide pools and in the subtidal zone; often found in the drift 55

54. **Branches thin, thread-like, mostly irregularly alternate, often with a tripartite tip;** tufts loose or compact, up to 5 cm high; brownish red to black; clear, apical cell visible at tip in young branches; found year round, from northern Massachusetts southward, in the mid to low intertidal zone, attached to rocks; also growing on mud, at edges of salt marshes .... *Gelidium pusillum* (Fig. 50)

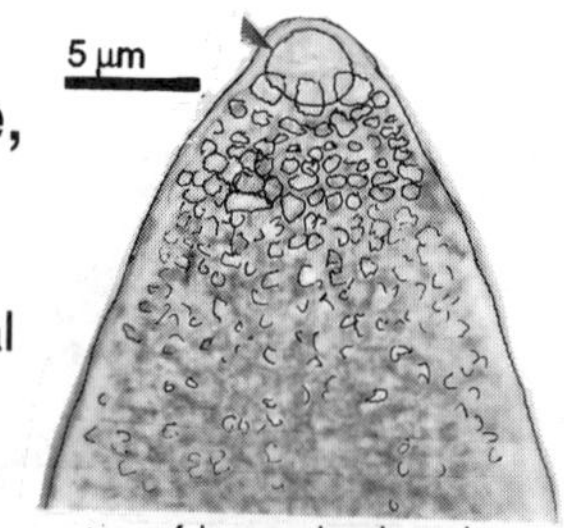

tip of branch showing single apical cell

54. **Branching dichotomous but frequently irregular or bunched at tips;** forming tight tufts, 2-4 cm high, almost black; no apical cell visible; uncommon; found year round, in southern Massachusetts and to the south, attached to rocks, in the mid to lower intertidal zone........ .... *Gymnogongrus griffithsiae* (Fig. 51)

55. **Branching very regularly dichotomous;** axes and branches rounded; thickness remaining constant throughout (up to 1.5 mm) but tapering at tips; thallus dark brownish red to almost black, 10-15 (20) cm high; found year round, on the open coast and in estuaries, in low intertidal pools, and in the subtidal zone, down to -18 m, attached to rocks; often found in the drift .... *Polyides rotundus* (Fig. 52)
55. **Branching originally dichotomous but becoming very irregular and entangled;** stiff tufts, with thin axes and branches less than 1 mm thick, with blunt tips; black, often yellowish or whitish at tips; found year round, on the open coast and in estuaries, in mid to low intertidal tide pools and in the subtidal zone, down to -10 m, attached to rocks; often found in the drift .... *Ahnfeltia plicata* (Fig. 53)

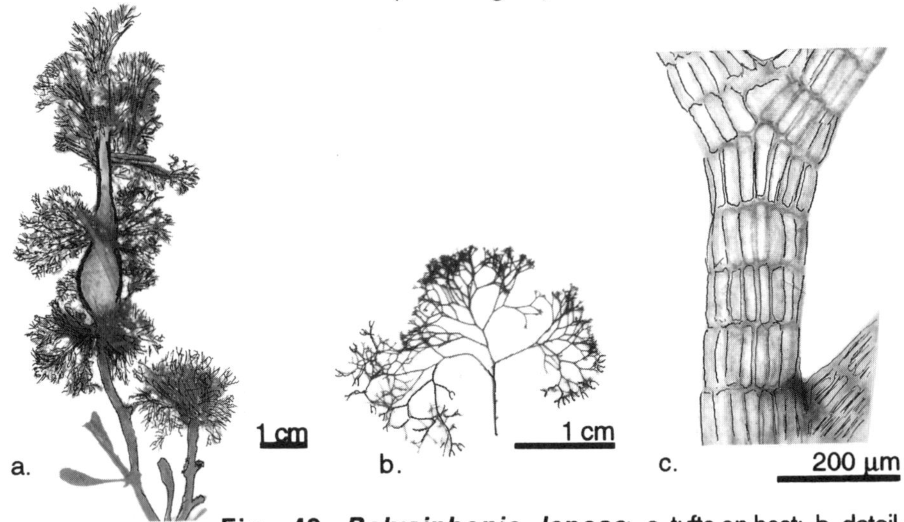

Fig. 49. ***Polysiphonia lanosa***; a. tufts on host; b. detail of branching; c. pericentral cells from surface view

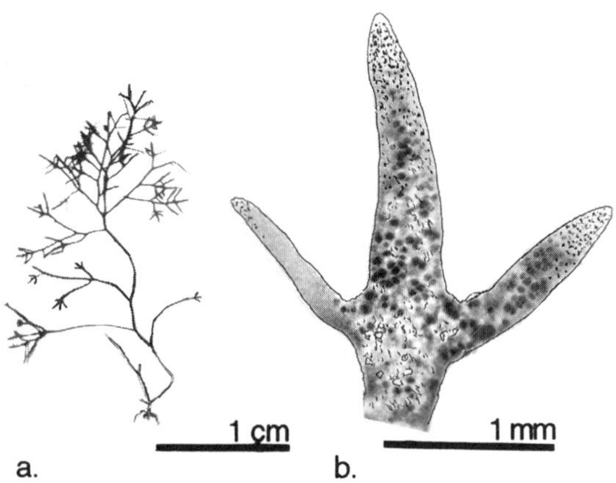

Fig. 50. ***Gelidium pusillum***
a. branching detail; b. tip showing typical branching and tetraspores

Fig. 51. ***Gymnogongrus griffithsiae***
individual specimens

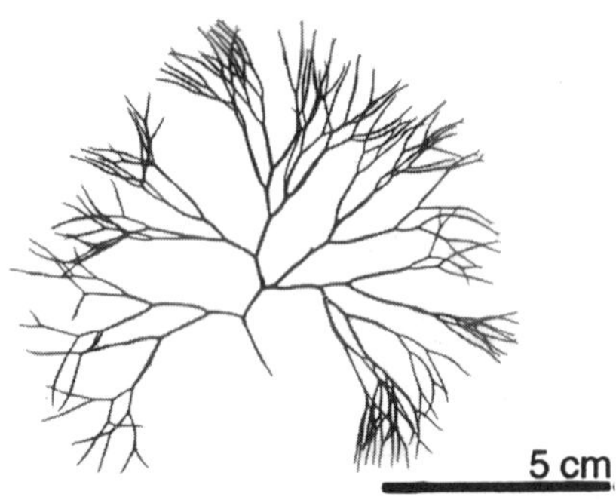

Fig. 52. ***Polyides rotundus***
specimen showing dichotomous branching

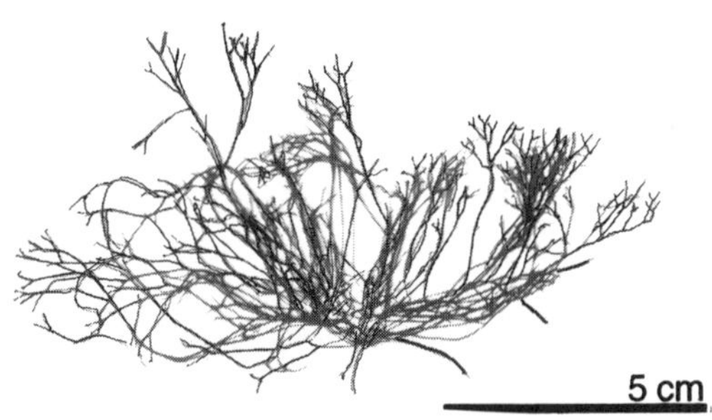

Fig. 53. ***Ahnfeltia plicata***
specimen showing irregular branching

56. **Slippery and worm-like, or at least soft and gelatinous** 57
56. **Not as above** 59

57. **Numerous, small, bushy branchlets, tapering at both ends**; thallus bushy, up to 40 cm long or more; 1 to several main axes, up to 5 mm wide, hollow at base, unbranched below, tapering at each end, with few large branches also tapered and profusely rebranched; pale brownish red to deep rosy red, sometimes yellowish; found sporadically from Connecticut northward, mostly on the open coast, in exposed areas, from early spring to early summer, attached to rocks, in deep tide pools, or in the shallow subtidal zone, down to -3 m .... *Gloiosiphonia capillaris* (Fig. 54)
57. **Branching sparse to moderate, with few secondary branches**.. 58

58. **Largest branches solid, less than 2 mm wide**; very soft, worm-like, elastic thallus, up to 20 cm long; branching sparse, irregular to almost dichotomous, with some rebranching; cylindrical but somewhat flattened and wider at branching points; light brownish or purplish red to burgundy red; found sporadically from Long Island Sound northward, on the open coast only, in the summer; forming slippery patches on rocks, particularly in exposed areas, in the mid to low intertidal zone, attached to rocks and shells .... *Nemalion helminthoides* (Fig. 55)
58. **Largest branches hollow and usually more than 2 mm wide;** (see p. 92, step 41, Fig. 38) .... *Dumontia contorta*

59. **Branched chains of regular barrel-shaped segments;** segments hollow, cylindrical, 1-2 mm wide, with regular diaphragms and slight constrictions at the diaphragm level; short (up to 8 cm tall), stubby tufts, stout and relatively coarse and crisp, with many tapering branches; pink to red, brownish or greenish; common from Cape Cod southward, particularly in estuaries, in summer and fall, in the subtidal zone, down to -12 m, attached to stones and shells, or epiphytic on eelgrass and other algae; often floating in the drift .... *Champia parvula* (Fig. 56)

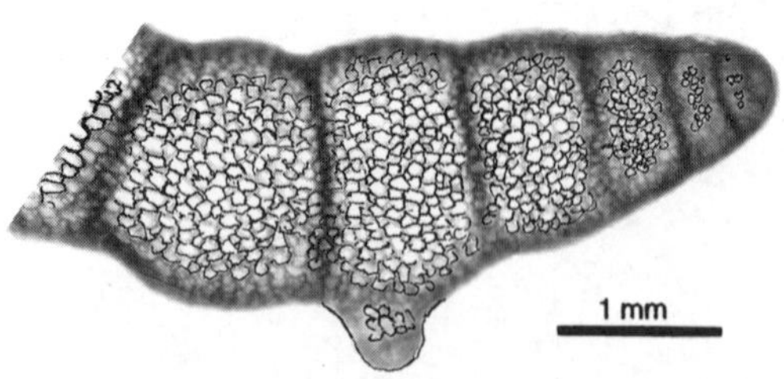

C. parvula, detail of branch tip showing segmentation and diaphragms

59. **Branches not as above** 60

60. **Branches long, sometimes twisted and hollow** (see p. 92, step 41, Fig. 38) .... *Dumontia contorta*
60. **Branches otherwise** 61

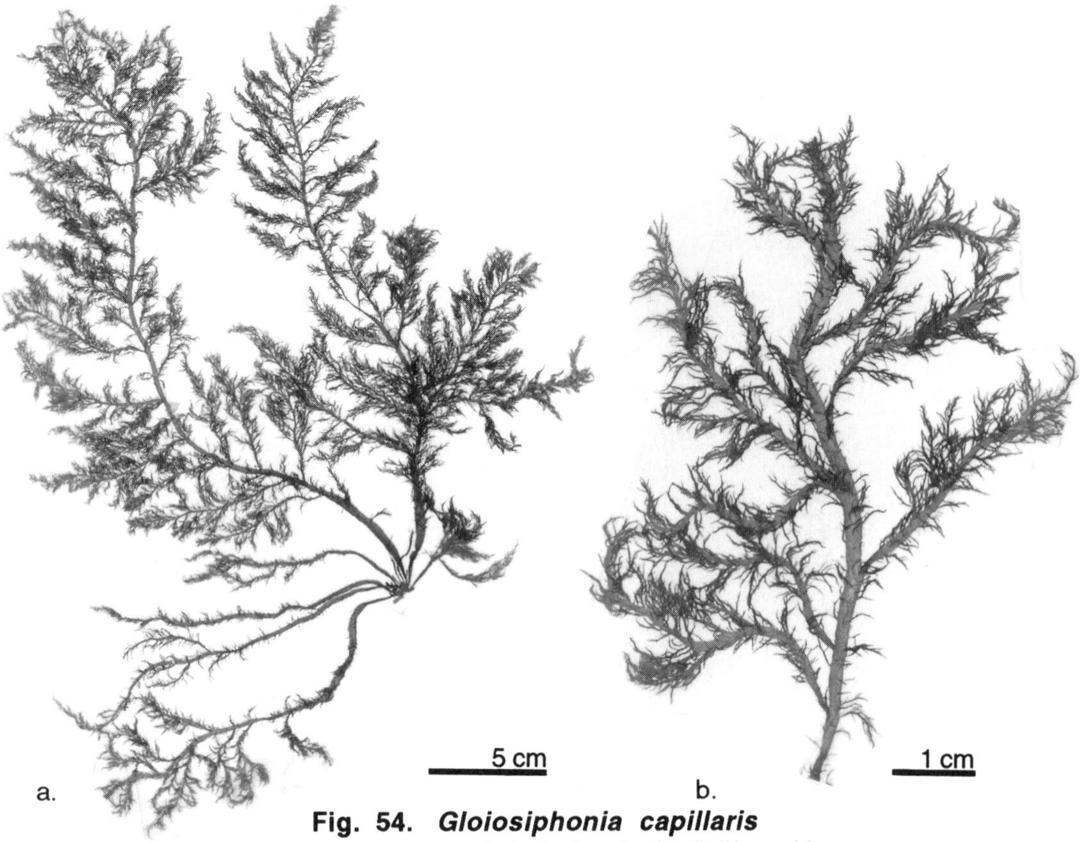

a. b.

Fig. 54. ***Gloiosiphonia capillaris***
a. whole thallus; b. detail of branching

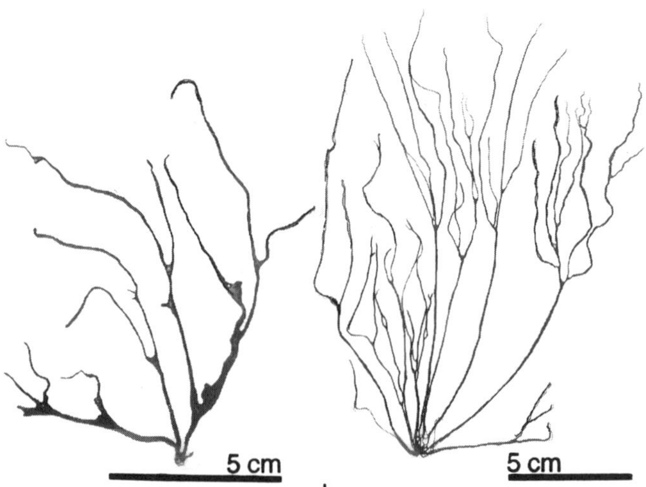

a. b.

Fig. 55. ***Nemalion helminthoides***
a. specimen with broad branching points
b. specimen with narrower branching points

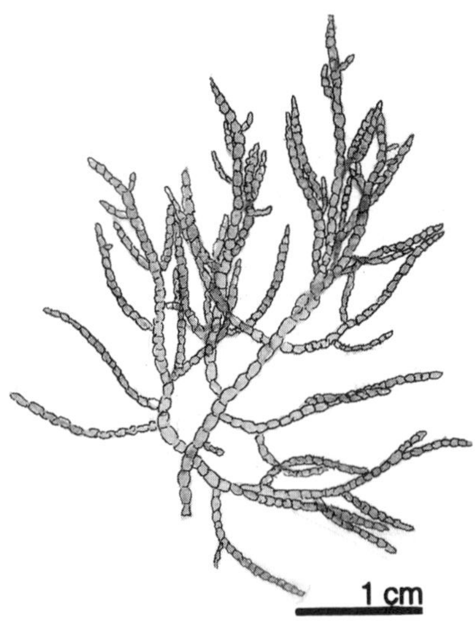

Fig. 56. ***Champia parvula***
branch with barrel-shaped segments

61. **Branches and branchlets unsegmented, hollow; constricted or tapering to base; terminal branchlets no more than 2 cm long;** thallus generally soft 62
61. **Branches otherwise** 66

62. **Terminal branches tapering to base, club to cigar-shaped,** often with a different color from the rest of the thallus; (*Chondria* spp. ♦3 of 4?) 63
62. **Terminal branches sharply constricted at base,** and with the same color as the axes; (*Lomentaria* spp. ♦2 of 3) 65

63. **Branchlets club-shaped, thick, fleshy and swollen, up to 2.5 mm wide, with a depressed tip** (drawing below); somewhat short (1-2 mm), very blunt, with tufts of hairs; thallus shaped like a broad Christmas tree, up to 10 cm long, dark brownish or purplish red, to rosy red or straw color; found in New Hampshire and to the south, in summer, at or just below the low tide level, mostly on the open coast, attached to rocks and shells, and occasionally epiphytic on other algae .... *Chondria dasyphylla* (Fig. 57)
63. **Branchlets spindle or cigar-shaped, with an emergent tip** 64

64. **Shaped like a narrow Christmas tree, with soft, fine branches, no more than 1 mm wide;** branchlets like a baseball bat, with somewhat blunt tips, 5 mm long (see below); thallus up to 15 (20) cm long, pale purplish pink to straw color; found in summer, in the subtidal zone of warm bays and estuaries, down to -1.5 m, attached to stones, or epiphytic on eelgrass and coarser algae, in quiet protected waters; often drifting on the bottom .... *Chondria baileyana* (Fig. 58)
64. **Branchlets (2-7 mm long), tapered at both ends, with abundant tufts of long hairs** (drawing below); axes not soft, up to 2.5 mm wide and up to 20 cm long; purplish red to straw color; found in northern Massachusetts and to the south, in summer, attached to rocks or shells, at or just below the low tide level, mostly on the open coast .... *Chondria capillaris* (Fig. 59)

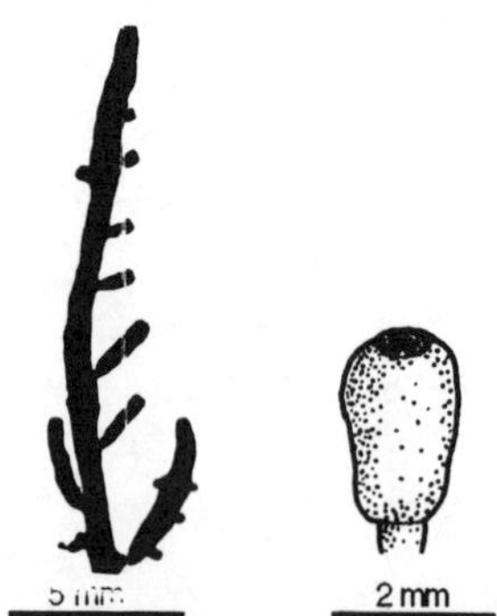

C. dasyphylla, branchlets

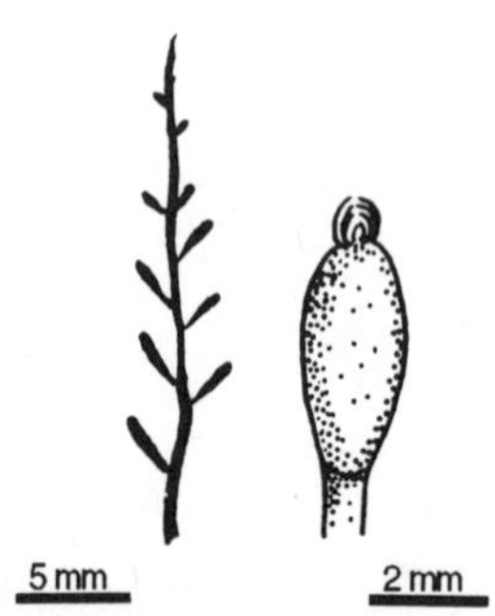

C. baileyana, branchlets

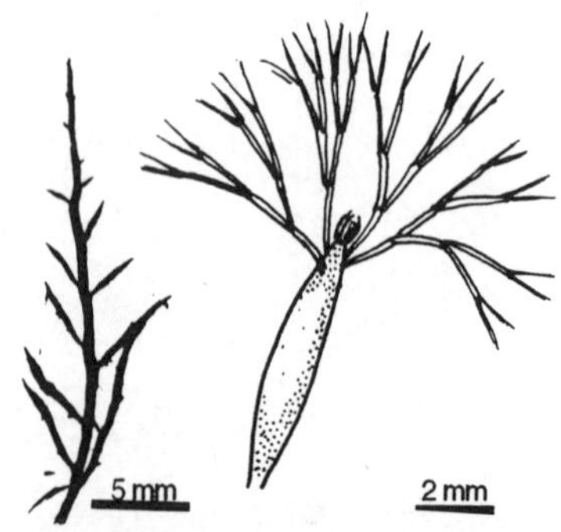

C. capillaris, branchlets

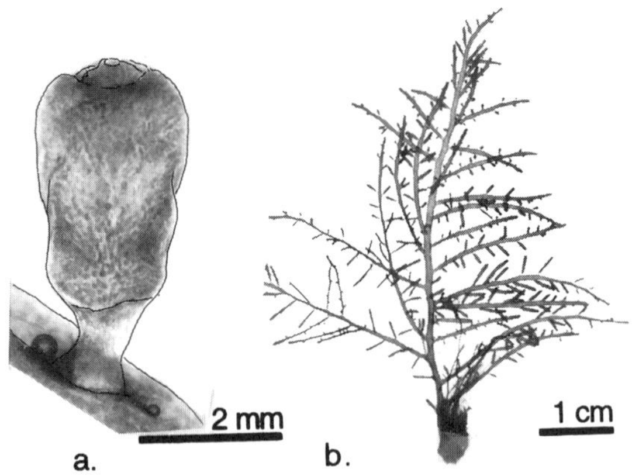

Fig. 57. ***Chondria dasyphylla,*** a. blunt, club-shaped branchlet; b. whole thallus

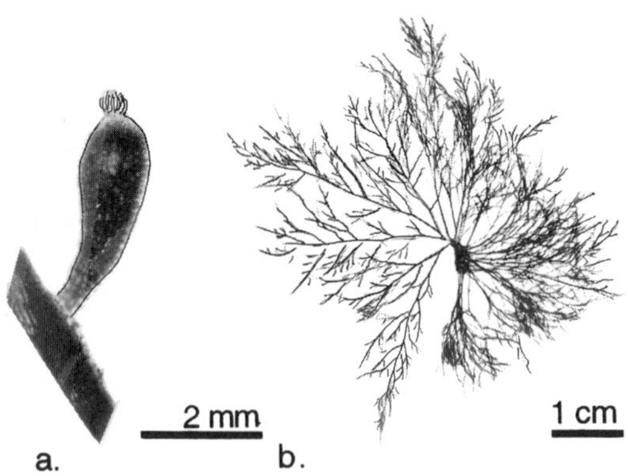

Fig. 58. ***Chondria baileyana;*** a. branchlet shaped like a baseball bat; b. whole thallus

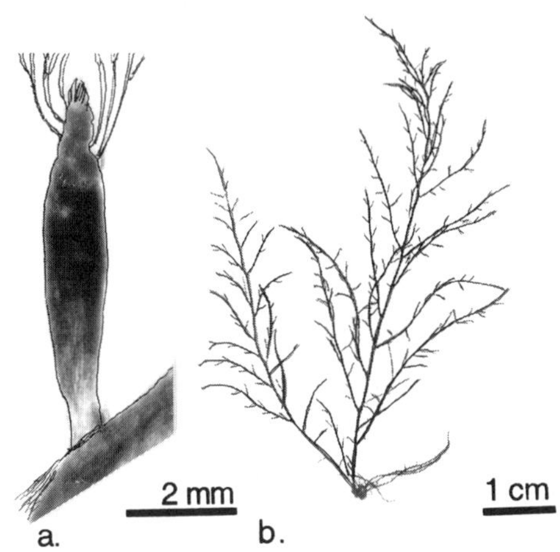

Fig. 59. ***Chondria capillaris***
a. cigar-shaped branchlet
b. whole thallus

65. **Branches tapered to end, often unilateral and curved**; soft tufts, up to 8 cm tall, with a few main axes; branches constricted at base and gently curved, hollow, except at base; pink to purplish or brownish red, occasionally yellowish; common in estuaries, from early summer to early winter; found in somewhat protected, shallow, subtidal zones, down to -5 m, attached to rocks and shells, or epiphytic on other algae and eelgrass; often free-floating .... *Lomentaria baileyana* (Fig. 60)

65. **Branches strait and tapered at both ends;** soft, highly branched, cylindrical axes, with short branchlets producing a bushy appearance; branchlets remaining red when dry; extremely variable; found year round, in the subtidal zone, down to -31 m, attached to rocks, or epiphytic on other algae; often free-floating; from Cape Cod northward, where it is restricted to estuaries .... *Lomentaria clavellosa* (Fig. 61)

66. **Tips of branches modified or with special appendages visible to the naked eye** (hairs, hooks, tendrils, pincers or bumpy proliferations) 67

66. **Visible appendages absent; tips of branches unmodified** ... 80

67. **Visible, fleshy hooks present, 0.4-1 cm long** 68

67. **Fleshy hooks absent** 69

68. **Lateral hooks scattered throughout** (see below); thallus bushy, delicate, up to 15 cm long, often in a Christmas tree pattern; fleshy branches covered with hair-like, short (2 mm), even branchlets; bright pinkish red, turning orange or straw color when dying; common year round, on the open coast and in estuaries, in the subtidal zone, down to -20 m or more, attached to rocks and shells, or epiphytic at the base of coarser algae; often found in the drift or entangled with other algae .... *Bonnemaisonia hamifera* (Fig. 62)

68. **Hooks restricted to tip of branches** (see below); thallus bushy, with a few irregular main branches; up to 45 cm long; fleshy branches covered with short (2-20 mm long), uneven branches, spur-like and darker than the branches; greenish to purplish red, gray or yellowish; uncommon; found in summer, from Cape Cod southward, in the quiet, warm waters of bays and estuaries, in the shallow subtidal zone, attached to rocks and shells; also found free-floating .... *Hypnea musciformis* (Fig. 63)

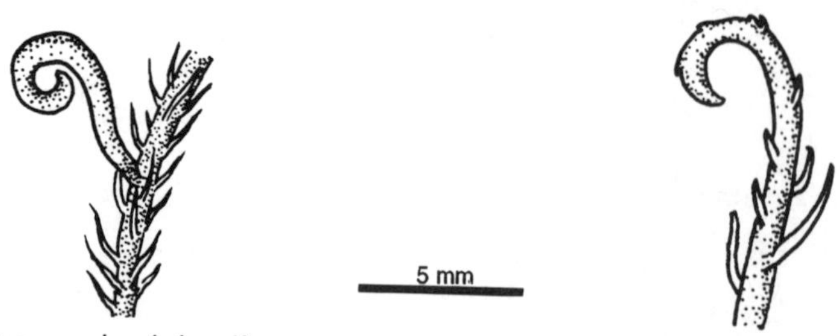

Bonnemaisonia hamifera *Hypnea musciformis*

details of hook

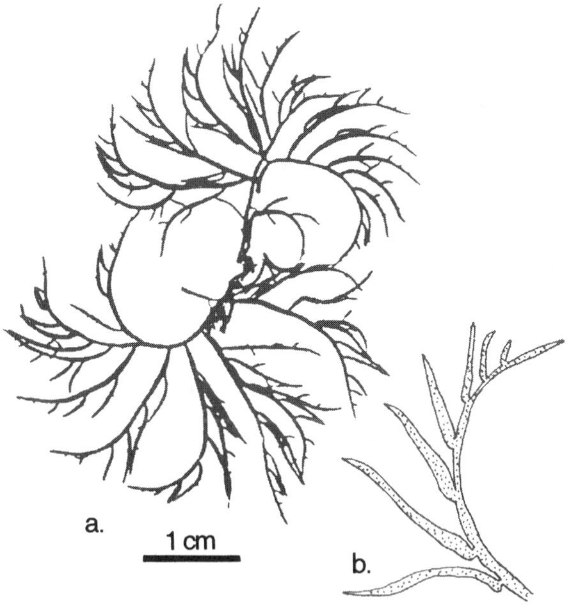

Fig. 60. ***Lomentaria baileyana***
a. tuft with curved branches and unilateral branching; b. detail showing constricted base of branches

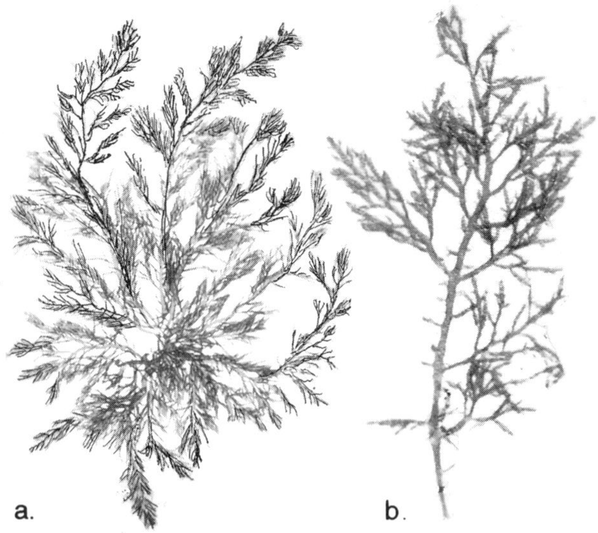

Fig. 61. ***Lomentaria clavellosa***
a. whole thallus
b. detail of branch

Fig. 62. ***Bonnemaisonia hamifera***
branch with lateral hooks

Fig. 63. ***Hypnea musciformis***
branch with terminal hooks

69. **Pigmented hairs or hair-like branchlets present;** hairs either covering the thallus or clustered in tufts 70
69. **Pigmented hairs absent or not obvious** 72

70. **Hair-like branchlets in clusters or tufts at tips of branches** (see p. 124, step 101, Fig. 90) .... *Rhodomela confervoides*
70. **Hairs not in tufts** 71

71. **Axes and branches thickly covered with a sheath of short, red hairs, 0.4-1.2 cm long** (hairs best seen in water); main axis solid, cylindrical, up to 75 cm long, with a few lateral branches, sparingly rebranched; branches often bearing small, dark, reproductive structures, rounded or elongated; thallus pinkish red to brownish or purplish red; older specimens may loose hairs; one of our most beautiful and spectacular species; common from spring to fall, and occasionally in winter, on the open coast and in estuaries, in the subtidal zone, down to -5 m, attached to rocks and shells, or epiphytic on other algae; often free-floating in the drift .... *Dasya baillouviana* (Fig. 64)
71. **Thallus with a delicate pale sheath of fine, short hairs (1-3 mm), except in older parts; branches with alternating bands of short and long cortication cells** (Fig. 65 b); thallus bushy, brittle, with many alternate branches; uniaxial construction with pericentral cells (p. 112, step 81) and cortication; pale or grayish pink to straw color; found from Cape Cod southward, in summer, in the shallow subtidal, in quiet, warm waters of bays and estuaries; attached to rocks and shells, but often forming free-floating masses .... *Spyridia filamentosa* (Fig. 65)

72. **Surfaces and/or margins with irregular proliferations (from small bumps to short, branched growths)** 73
72. **Irregular proliferations absent** 74

73. **Small, irregular, rounded or elongated proliferations or bumps, present mostly on the surface of blades** (see p. 96, step 48, Fig. 45) .... *Mastocarpus stellatus*
73. **Elongated, branched proliferations, present mostly along the margins of blades** (see p. 96, step 48, Fig. 46) .... *Fimbrifolium dichotomum*

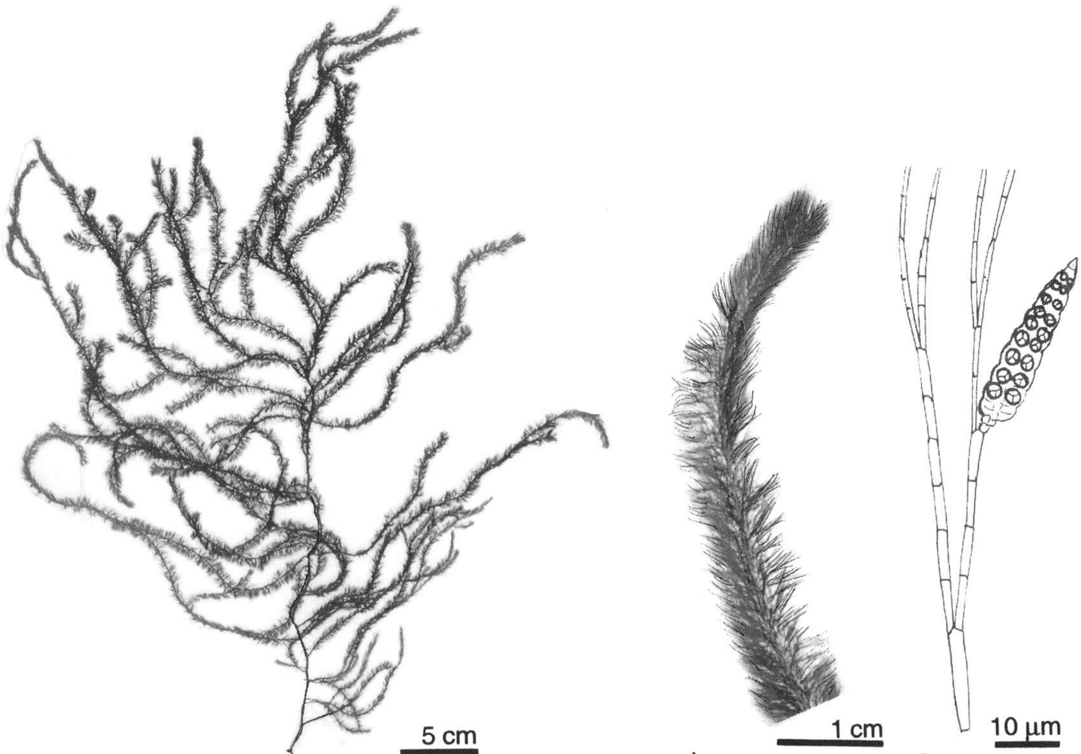

a. b. c.

Fig. 64. ***Dasya baillouviana***
a. whole specimen; b. detail of hairs
c. hair with tetrasporangia

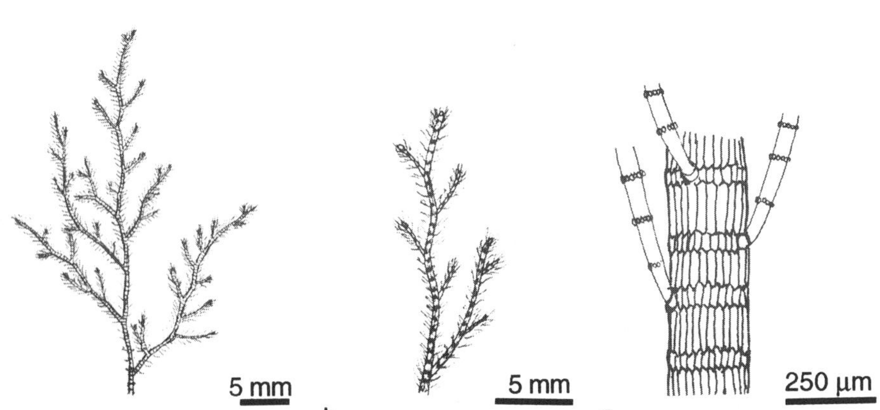

a. b. c.

Fig. 65. ***Spyridia filamentosa***
a. branch; b. branch showing
hairs and striations; c. detail of cortical striations

74. **Coiled, grapevine-like tendrils at tips of branches**; thallus bushy, firm and fleshy, up to 50 cm long, with a distinct main axis up to 2 mm wide, broader than the branches; abundant, irregular, lateral branching, with smaller branches tapering at both ends; often covered with dark reproductive lumps; light red to dark purplish red, but also yellowish or greenish; very common and variable; common year round, on the open coast and in estuaries, in the lower intertidal and subtidal zones, down to -10 m or more, attached to rocks and epiphytic on other algae; often in the drift (see also page 134, step 117, Fig. 103) .... *Cystoclonium purpureum* var. *cirrhosum* (Fig. 66)

74. **Pincers present at tips of branches** (tips forked inward like a lobster claw) 75

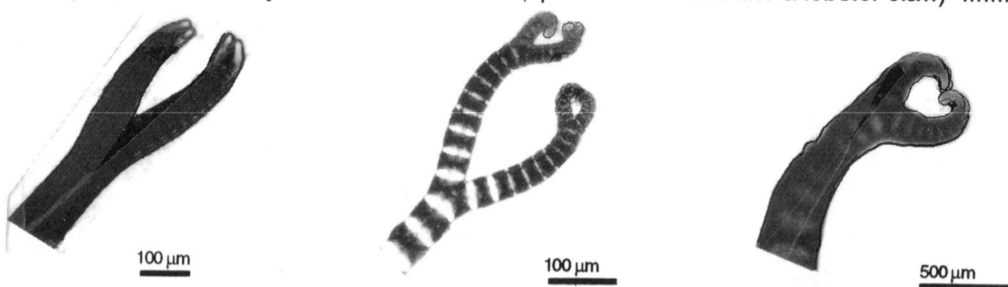

P. lanosa *C. cimbricum* *C. nodulosum*

tips of branches with pincers

75. **Black tufts, epiphytic on *Ascophyllum nodosum*** (see p. 98, step 51, Fig.49) .... *Polysiphonia lanosa*

75. **Branches appearing beaded or banded by cortication** (illustrated below); (*Ceramium* spp. ♦4 of 6?) 76

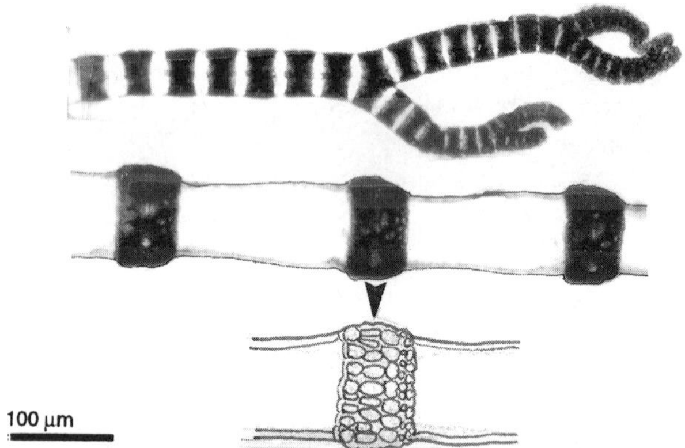

C. cimbricum, partial cortication (covering nodes only)
upper and lower sections of thallus

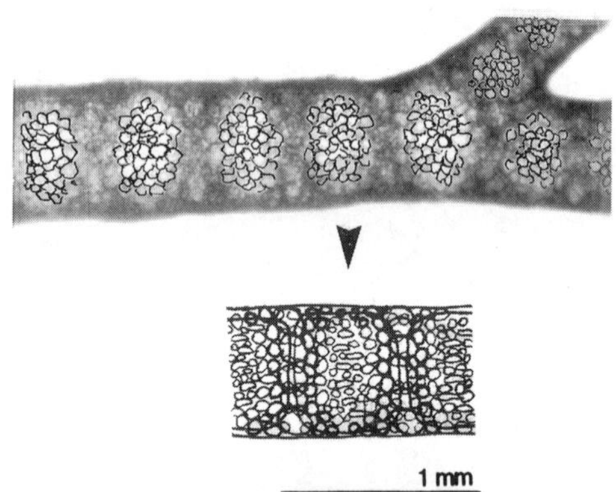

C. nodulosum, complete cortication

76. **Cortication limited to the nodes, producing a distinct beaded appearance of clear internodes, alternating with pigmented, dark nodes** 77

76. **Cortication covering most of the thallus; beaded appearance observable, but often obscured;** bushy thallus, up to 40 cm tall, dichotomous branching abundant; pincers may disappear in older specimens; texture fine to coarse; color yellowish to brownish, dark red or pink; color, shape and texture extremely variable; very common year round, in the intertidal and subtidal zones, attached to stones and shells, or epiphytic on other algae such as *Codium fragile*; often found in the drift (see also page 132, step 115, Fig. 100) .... *Ceramium nodulosum* (Fig. 67)

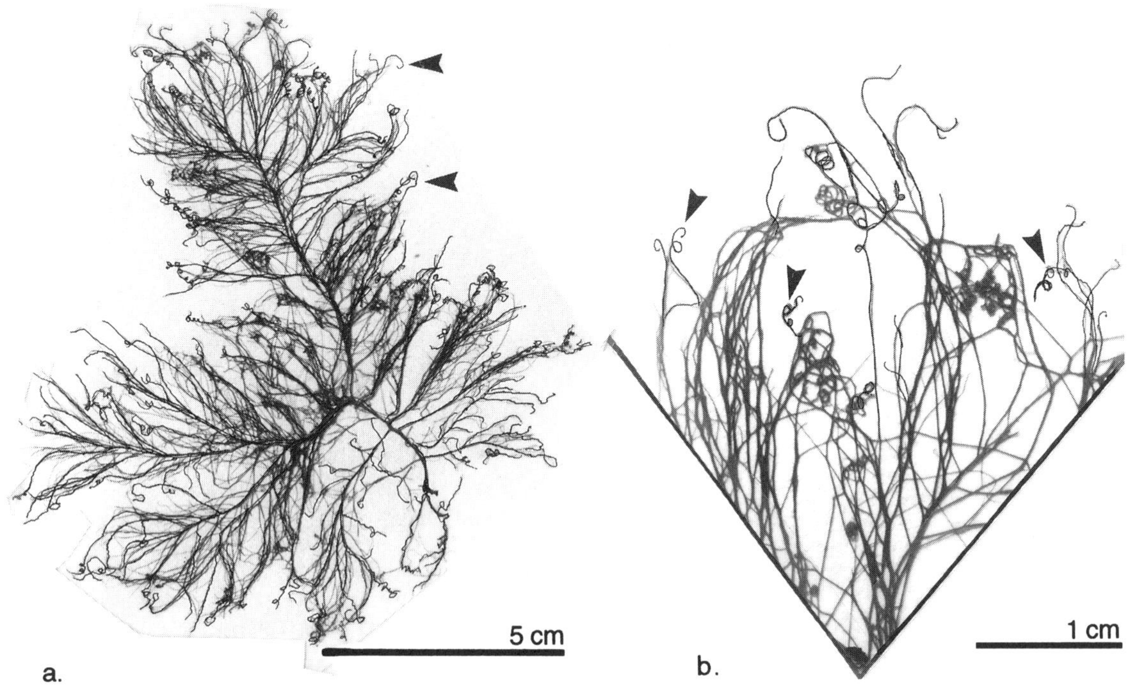

Fig. 66. ***Cystoclonium purpureum*** **var.** ***cirrhosum***
a. whole thallus; b. detail of branch tips with tendrils

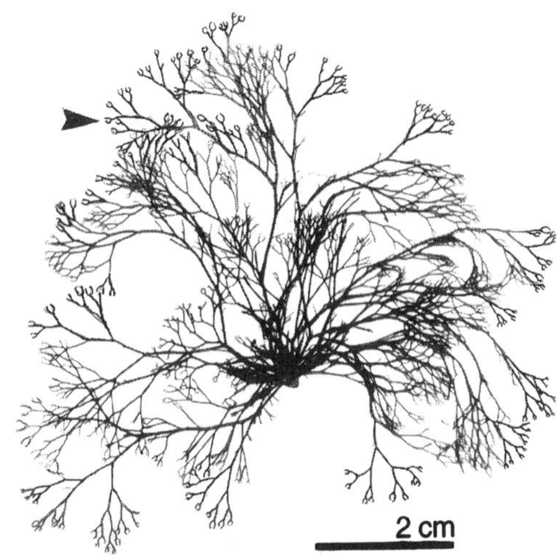

Fig. 67. ***Ceramium nodulosum***
single specimen showing pincer tips

77. **Mature axes with long, clear internodes (more than 3 times the length of the adjacent, dark nodes)** 78

77. **Mature axes with short, clear internodes (less than 3 times the length of the adjacent, dark nodes);** tetrasporangia protruding from node, or completely emergent; tufts dark purplish to brownish red, up to 6 cm long, with a matted growth; axes with numerous short branchlets; found year round, on the open coast and in estuaries, attached to rocks or shells, and epiphytic on crustose corallines, in the mid to low intertidal zone; often found under fucoids .... *Ceramium deslongchampii* (Fig. 68)

78. **Tetrasporangia emergent from nodal cells, or naked; nodal cells in 2-6 regular, transversal rows; internodes up to 8 times as long as node;** thallus delicate but brittle, up to 10 cm long, with dichotomous branching; rosy red to purplish or brownish red; found in sheltered bays and estuaries, in the shallow subtidal zone, mostly epiphytic on eelgrass or other algae, but also attached to rocks and shells .... *Ceramium cimbricum* (Fig. 69)

78. **Tetrasporangia completely immersed or covered by at least a few nodal cells; nodal cells irregularly arranged** 79

79. **Short, lateral branchlets abundant, often forked; dark nodes slightly swollen;** rosy red to brownish red tufts, with dichotomous and spreading branches, up to 20 cm long; found year round, but common only from spring to fall, in warm sheltered bays and estuaries, in the shallow subtidal zone, attached to rocks and shells, or epiphytic on eelgrass and other algae .... *Ceramium diaphanum* * (Fig. 70)

79. **Lateral branchlets absent or few; nodes not swollen;** rosy red to brownish red tufts, up to 8 cm long, with dichotomous branching; tetrasporangia covered by nodal cells; found year round, but common only from spring to fall, in warm sheltered bays and estuaries, in the shallow subtidal zone, attached to rocks and shells, or epiphytic on eelgrass and other algae; common on muddy bottoms .... *Ceramium strictum* * (Fig. 71)

* *C. diaphanum* and *C. strictum* are difficult to separate and their taxonomy is in transition. Many authors, such as Bird and McLachlan (1992) recognize them as belonging to the same complex and refer to them as the *C. diaphanum-strictum* complex.

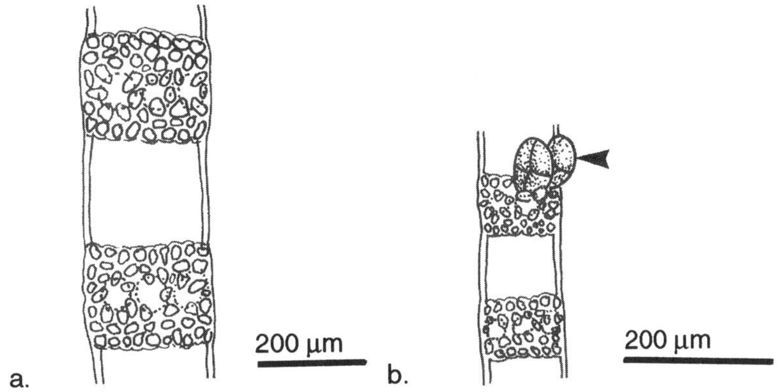

Fig. 68. ***Ceramium deslongchampii***
a. portion of filament showing short internode
b. nodes showing emergent tetrasporangia

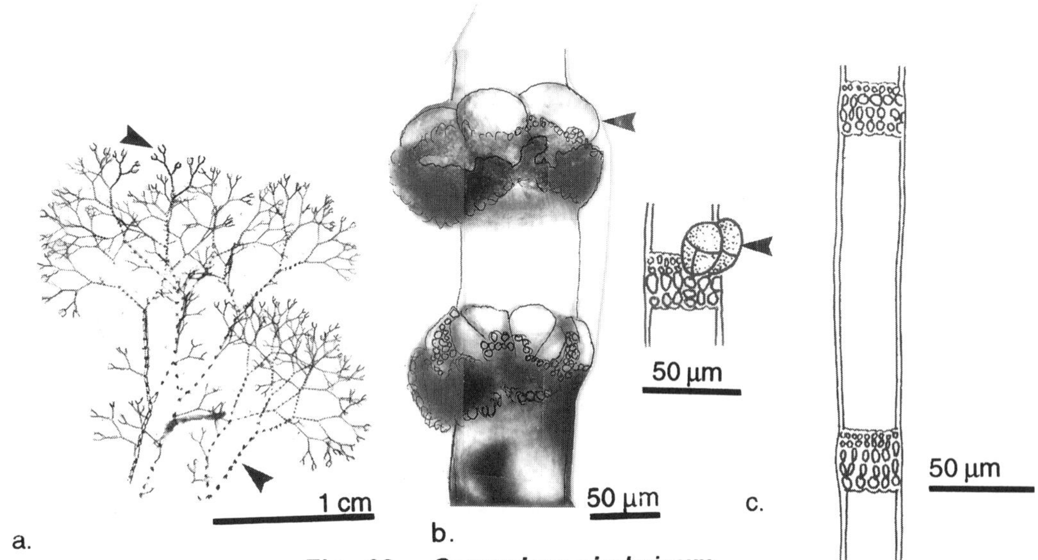

Fig. 69. ***Ceramium cimbricum***
a. branches showing banding and pincer tips;
b. nodes with emergent tetraspores; c. detail of nodes with regular rows of cells

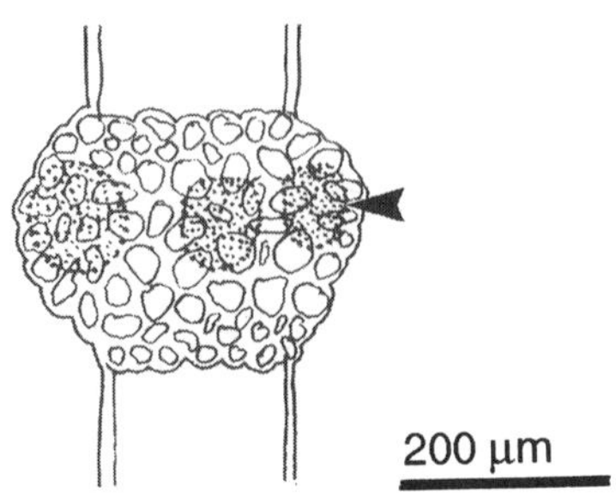

Fig. 70. ***Ceramium diaphanum***
detail of node with immersed sporangia

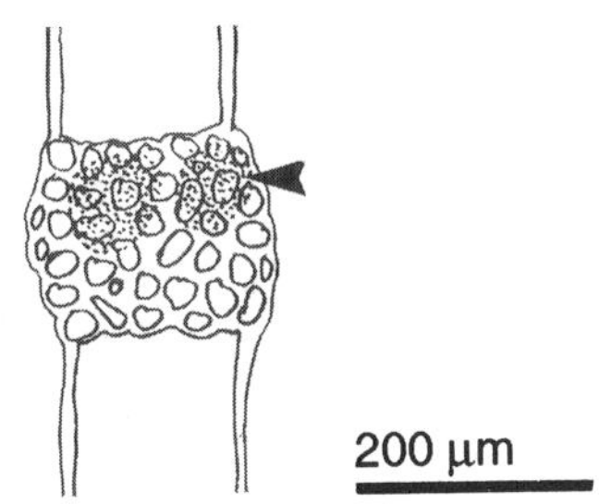

Fig. 71. ***Ceramium strictum***
detail of node with immersed sporangia

80. **Thallus very thin and hair-like, at least in upper parts** (microscopic characters needed for identification) 81
80. **Thallus not hair-like; axes fleshy and cylindrical** 114

81. **Filament very fine, uniseriate** (chain of cells in a single row, attached end to end; drawing below), hair-like, soft, without pericentral cells (drawing below), with or without cortication (p. 108, step 75), with or without branches 82
81. **Filament polysiphonous** (chain of axial cells, each surrounded by specific numbers of pericentral cells; drawing below); with or without a surface layer of cortication; remaining hair-like, at least in upper parts 100

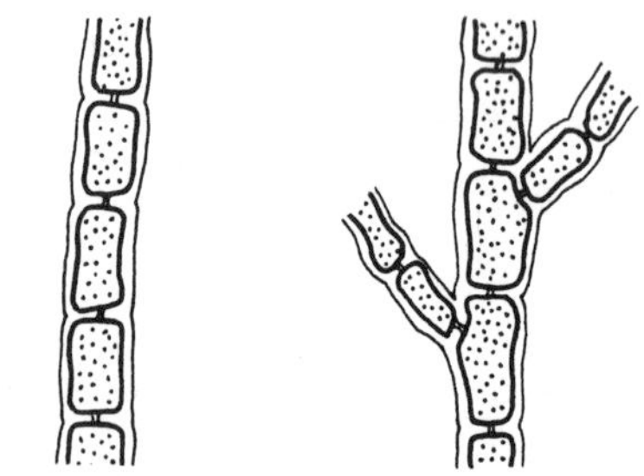

unbranched   branched

Uniseriate filament

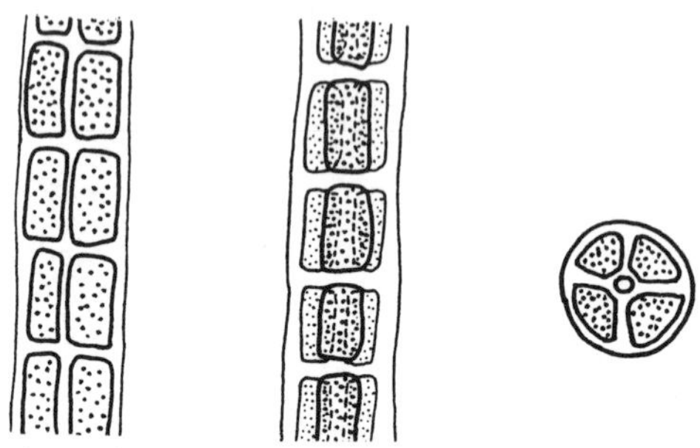

surface view   side view   cross section

Polysiphonous filament
with tiers of 4 pericentral cells

82. **Filament unbranched** 83
82. **Filament branched** 84

83. **Filament up to 2 cm long, epiphytic on other algae or on eelgrass,** mostly uniseriate, but occasionally redividing longitudinally in upper parts; monospores in cells with one oblique wall; found year round, on the open coast and in estuaries, in intertidal and shallow subtidal zones, in quiet waters .... *Erythrotrichia carnea* (Fig. 72)
83. **Filaments up to 15 cm long, mostly epilithic (attached to rocks);** uniseriate when young, becoming pluriseriate with age; cells usually flattened; soft and slippery; pale yellow to brown, purplish, red or black; monospores in cells without oblique walls; common in spring and early summer, on the open coast and in estuaries, in exposed areas, in the mid to high intertidal zone, forming large slippery patches on rocks and wood .... *Bangia atropurpurea* (Fig. 73)

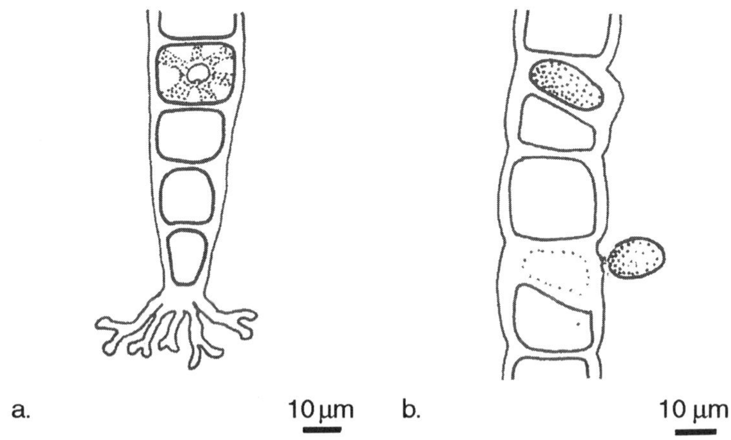

Fig. 72. ***Erythrotrichia carnea***
a. base of filament; b. portion of filament with monospores in cells with oblique walls

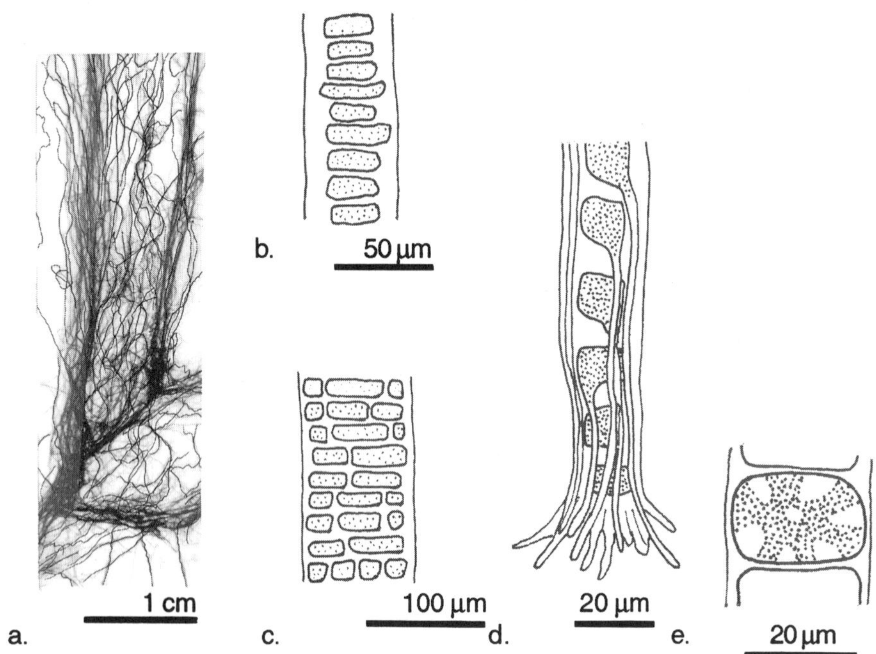

Fig. 73. ***Bangia atropurpurea***
a. cluster of filaments; b. young uniseriate filament; c. older , pluriseriate filament; d. base of filament showing descending rhizoids; e. cell with plastid and pyrenoid

84. **Very small, fuzzy growth of branched filaments, less than 1 cm high** 85
84. **Thallus longer than 1 cm** 86

85. **Cells in a thick gelatinous sheath, often widely spaced;** branching irregular; filaments up to 6 cm long, isolated or in clusters; pinkish red; found in summer and fall, in quiet waters of estuaries or bays, in the subtidal zone, down to -12 m, attached to hydroids, or epiphytic on eelgrass and other algae .... *Stylonema alsidii* (Fig. 74)
85. **Cells linked by pit-connections (drawing below); thick sheath absent;** branching alternate, often unilateral; hairs may be present; tufts pink to dark red, often clustered and turf-like; found year round, in the intertidal and subtidal zones, growing on, or within hydroids, eelgrass and other algae; also found on rocks; (♦14 local species or more, difficult to separate for the non specialist; taxonomy is in transition) .... *Audouinella* spp. (Fig. 75)

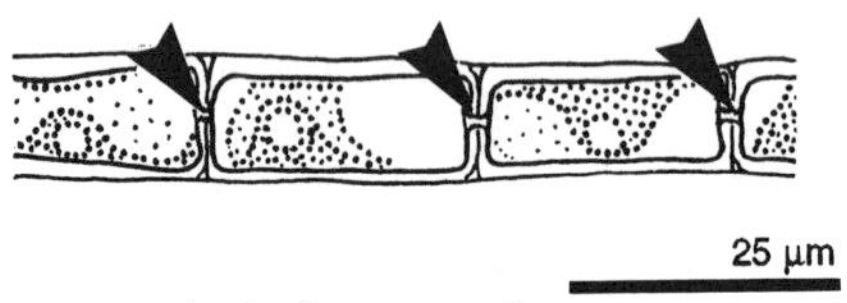

typical pit-connections

86. **Branches appearing beaded or banded by cortication; with pincers usually present at tips;** (p. 108, step 76, Figs. 67-71) .... *Ceramium* spp.
86. **Cortication if present, not banded; pincers absent** 87

87. **Cells very large (up to 1.5 mm wide), somewhat balloon or pear-shaped, swollen at tips, contracted at base, clearly visible with the naked eye;** thallus bushy, up to 8 cm long, branching appearing dichotomous; light pinkish to red; found from Cape Cod southward, in summer and early fall, in the warm waters of bays and estuaries, in the shallow subtidal zone, down to -5 m, attached to stones, shells and wood. This species has not been recorded for New Hampshire or Maine, although it occurs in the Gulf of St. Lawrence .... *Griffithsia globulifera* (Fig. 76)
87. **Cells not visible to the naked eye** 88

88. **Lateral branches mostly opposite or whorled; main branches often alternate** (branch and branchlet pattern for each species is different and distinctive) 89
88. **Branching alternate or irregular throughout** 93

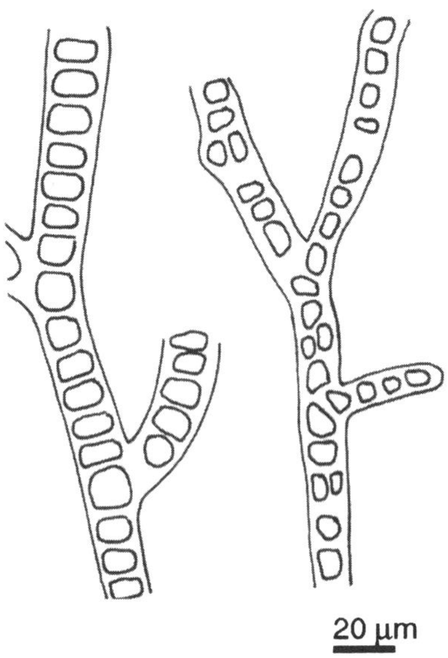

Fig. 74. ***Stylonema alsidii*** branch detail

Fig. 75. ***Audouinella* sp.** branching detail with monospores

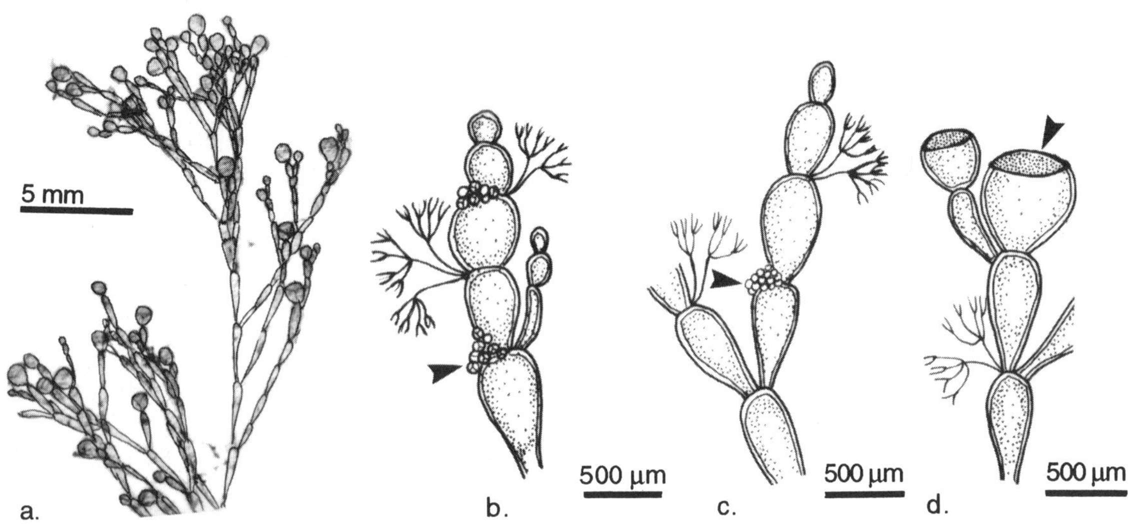

Fig. 76. ***Griffithsia globulifera***; a. portion of thallus; b. tetrasporophyte with tetraspores; c. female gametophyte with carposporophytes; d. male gametophyte with spermatangia

89. **Basal cell of branches and sometimes branchlets, shorter than the cell above it; gland cell if present, adjacent to 2-3 vegetative cells** 90

89. **Basal cell of branches and branchlets unmodified; gland cell if present, adjacent to 1 vegetative cell only** 92

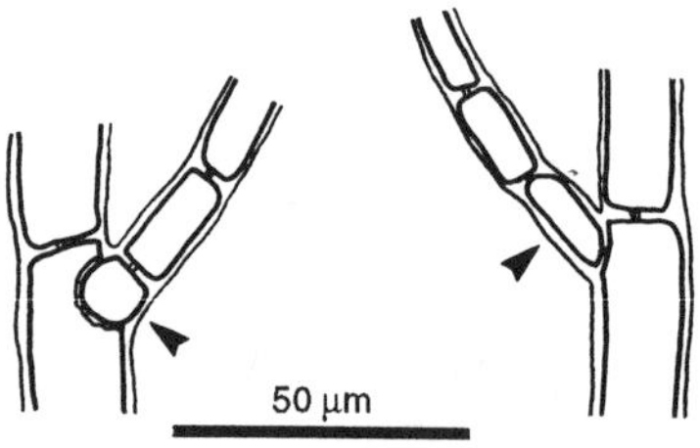

short basal cell   unmodified basal cell

basal cell of branch

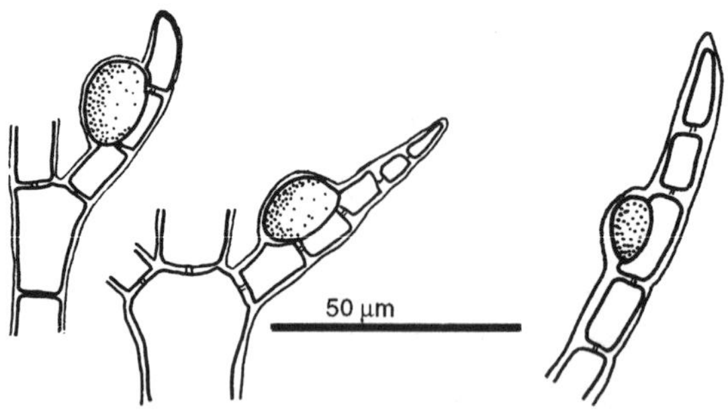

adjacent to 2-3 vegetative cells   adjacent to 1 vegetative cell

gland cells

90. **Axial cells each bearing regular pairs of short, undivided and opposite branchlets which taper abruptly; branching in a flat plane; gland cells absent in New England specimens;** tufts soft and feather-like, rosy pink, up to 10 cm long; descending rhizoids present near base; found year round, from northern Massachusetts to the north, on the open coast and in estuaries, in low tide pools and in the subtidal zone, down to -35 m, attached to rocks and epiphytic on other algae .... *Antithamnionella floccosa* (Fig. 77)

90. **Branching not as above; gland cells usually present** 91

91. **Opposite pairs of branches in 4 longitudinal planes, with successive pairs at right angles from each other; branching crowded and tufted at tips; if present, gland cells attached to a 3 to 4-celled spur-branchlet;** dense tufts, soft and delicate, up to 10 cm long, pinkish red; branchlets in 2 opposite rows or unilateral; very common; found year round, on the open coast and in estuaries, in the intertidal or subtidal zones, down to -12 m, attached to rocks, shells and mud, or epiphytic on eelgrass or other algae .... *Antithamnion cruciatum* (Fig. 78)

91. **Branches very regular, in 2 planes; tips of branches crowded, but flattened; rhizoids often present near tips of branchlets; gland cells abundant, attached to a branchlet up to 7 cells long;** small, soft tuft, bright pink, up to 3 cm long; recently introduced to southern New England, but very common year round, from Massachusetts to Connecticut; found on the open coast and in estuaries, in the low intertidal and upper subtidal zones, epiphytic on *Chondrus crispus, Corallina officinalis, Ulva lactuca* and other algae, as well as attached to rocks. Only sterile specimens have been observed in our area .... *Antithamnion pectinatum* (Fig. 79)

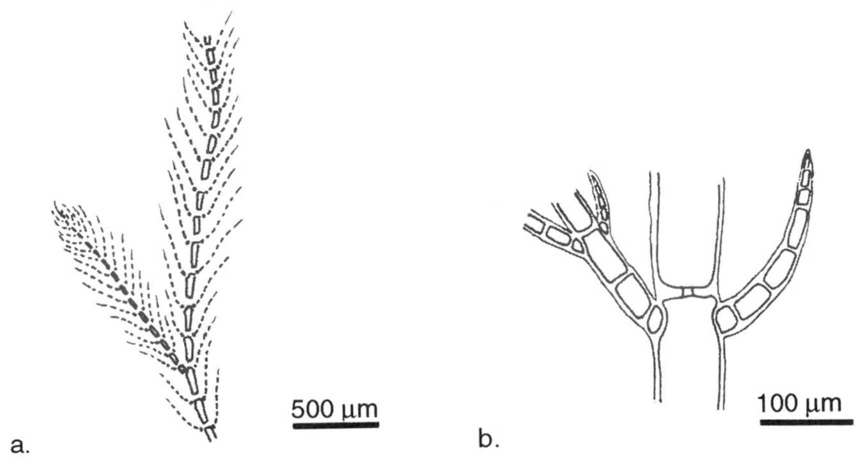

Fig. 77. *Antithamnionella floccosa*
a. branching pattern; b. detail of opposite branchlets

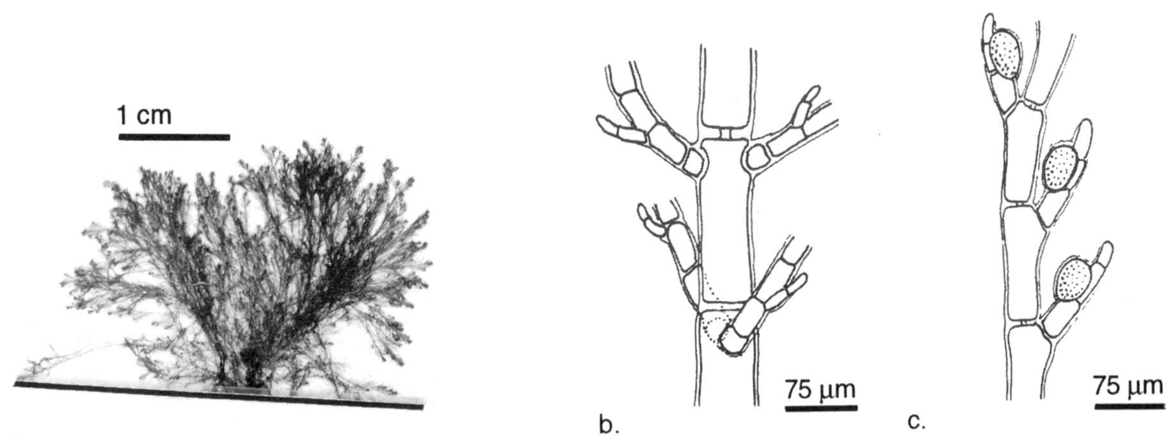

Fig. 78. *Antithamnion cruciatum*
a. tuft; b. detail of branching pattern with successive opposite pairs of branches at right angle to each other; c. branchlets with gland cells

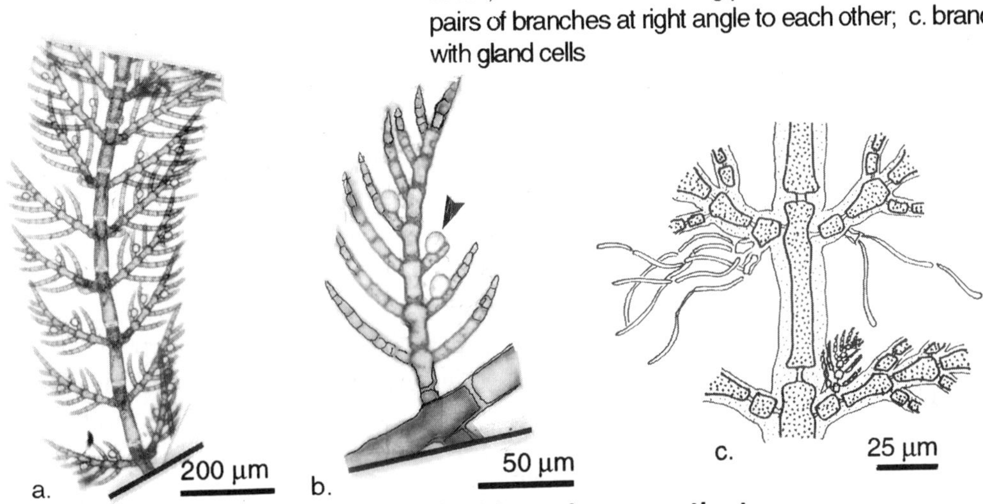

Fig. 79. *Antithamnion pectinatum*
a. detail of branching pattern; b. branchlet with gland cell
c. detail with rhizoids and miniature branchlet

92. **Branching flattened in one plane; with opposite or whorled, recurved branches bearing short branchlets on their upper side only; tips somewhat pincer shaped;** tufts delicate but somewhat stiff, rosy red, up to 12 cm long; found year round, on the open coast and in estuaries, in the low intertidal and subtidal zones, down to -12 m, attached to rocks, wood, shells and mud .... *Pterothamnion plumula* (Fig. 80)

92. **Axial cells bearing 2-4 opposite or whorled, irregularly branched branchlets;** tufts delicate, rosy pink, up to 10 cm long; common year round, from Cape Cod northward; uncommon to the south; found on the open coast and estuaries, in low tide pools and in the subtidal zone, down to -40 m, usually epiphytic on other algae .... *Scagelia pylaisaei* (Fig. 81)

93. **Cortication absent; horizontal filaments with unicellular rhizoids in the shape of a small holdfast;** often detached and looking like red cotton balls 94

93. **Basal axes variously corticated or with descending, multicellular, rhizoidal filaments from basal cells** 95

94. **Gland cells at nodal intersections of 2 cells**; dense, matted, and rounded tufts, up to 3 cm high, deep rose red or red; branching irregular or alternate; axes not distinct; basal filaments producing rhizoidal holdfasts for attachment; found from late witer to late fall, on the open coast and in estuaries, in the low intertidal and subtidal zones, down to -20 m, epiphytic on other algae; often found floating . This tuft is the tetrasporophytic phase of *Bonnemaisonia hamifera* and is no longer considered a valid taxon .... "*Trailliella intricata*" (Fig. 82)

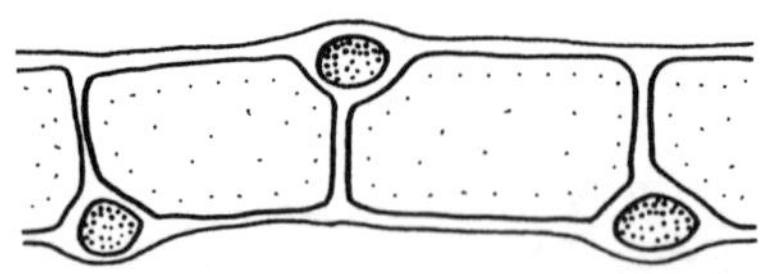

gland cells at nodal intersection of 2 vegetative cells

94. **Gland cells absent;** dense, round tufts, pinkish red to bright red, 1-4 cm in diameter; without a distinct main axis; branching near tips usually alternate or unilateral, but opposite branching also found on same individual; rhizoidal holdfasts on basal filaments; very common south of Cape Cod, but occurs only sporadically to the north; found year round, on the open coast and in estuaries, in the low intertidal and shallow subtidal zones, down to -12 m, attached to rocks and mud or epiphytic on other algae; often floating and accumulating in large masses in the drift .... *Spermothamnion repens* (Fig. 83)

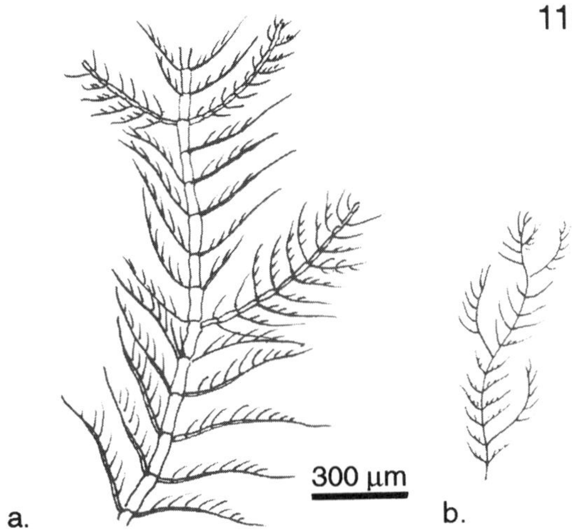

Fig. 80. ***Pterothamnion plumula***
a. detail of branching pattern
b. branching pattern at tip

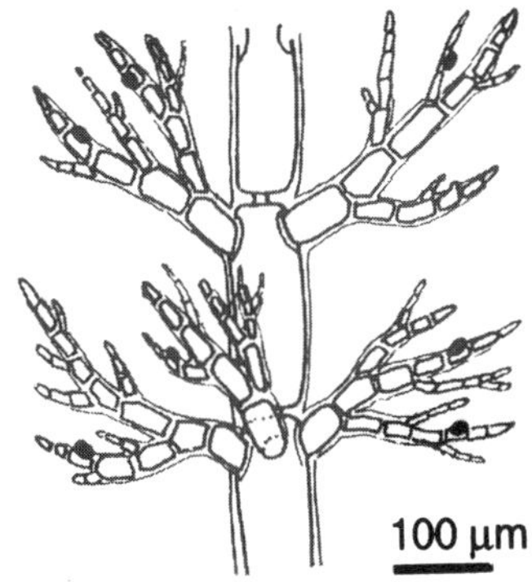

Fig. 81. ***Scagelia pylaisaei***
detail of branching pattern

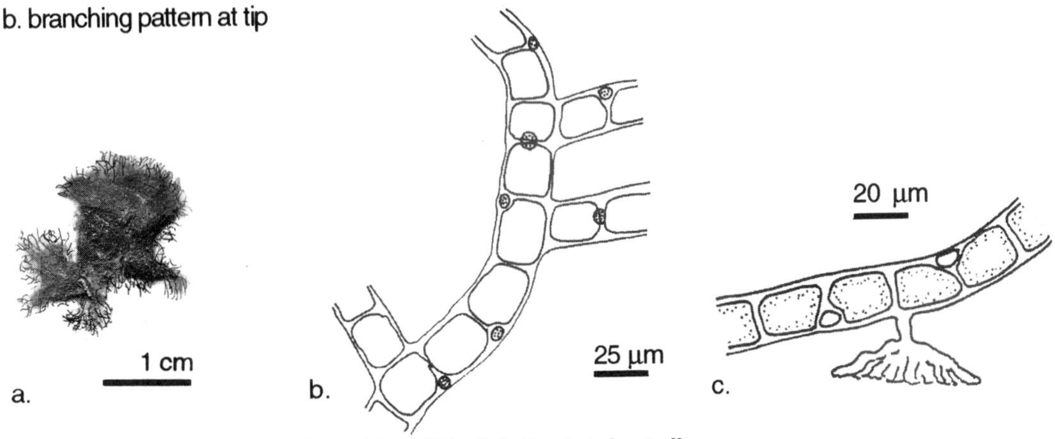

Fig. 82. ***"Trailliella intricata"***
a. tuft; b. detail of branching pattern; c. rhizoidal cell

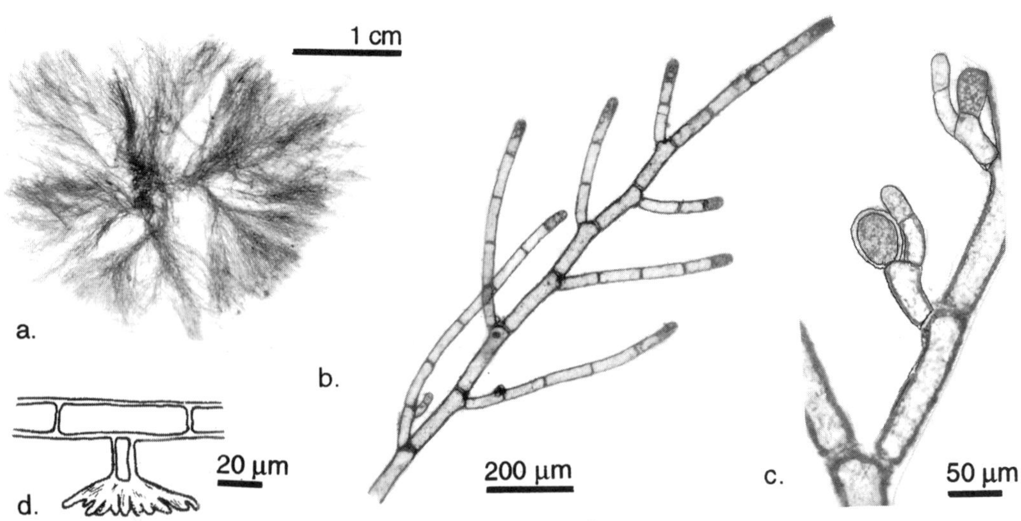

Fig. 83. ***Spermothamnion repens***
a. tuft; b. detail of branch tip; c. tetrasporangium
d. rhizoidal cell

95. **Tufts of spores in chain-like rows** (seirospores; drawing below); thallus soft, delicate, often with a triangular outline, up to 10 cm high, rosy pink to red; uncommon, particularly north of Cape Cod, found during the summer only, in warm bays and estuaries, in the subtidal zone, down to -10 m, attached to wood pilings, or epiphytic on eelgrass and other algae. Difficult to distinguish from some species of *Callithamnion* when not reproductive .... *Seirospora interrupta* (Fig. 84)

95. **Spores not in chain-like rows** 96

96. **More than 4 spores per cluster** (polyspores; drawing below); very delicate, bushy tufts, up to 5 cm tall, pinkish; branches forming a flat feather-like triangle with a naked stalk corticated at base only; found from Cape Cod southward, only during the summer, in the subtidal zone, attached to rocks and wood, or epiphytic on eelgrass and other algae. Difficult to distinguish from some species of *Callithamnion* when not reproductive .... *Pleonosporium borreri* (Fig. 85)

96. **Sporangia producing 4 spores** (tetraspores; drawing below); deep pink, red to brownish, up to 10 cm long 97

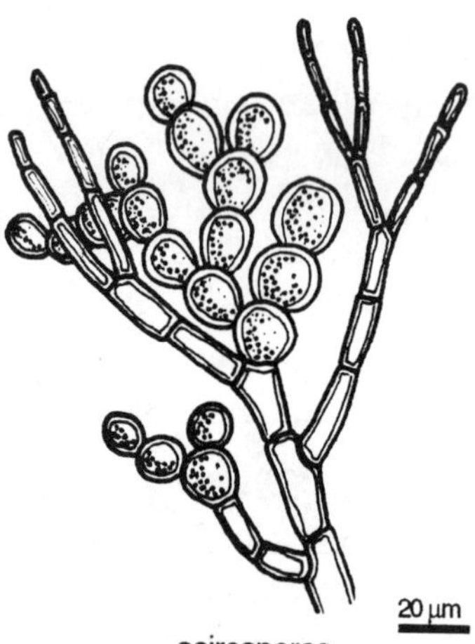

seirospores

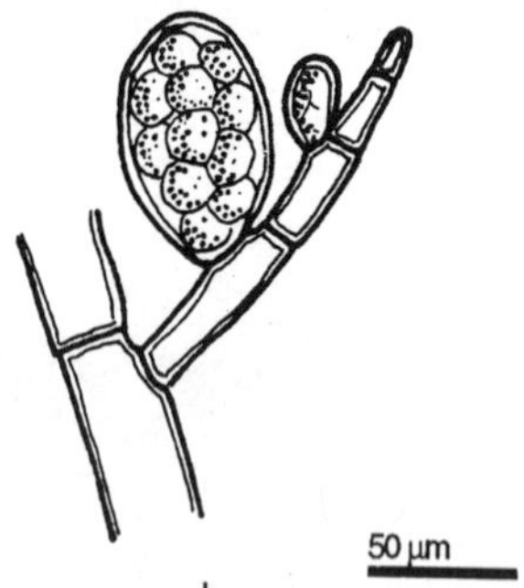

polyspores

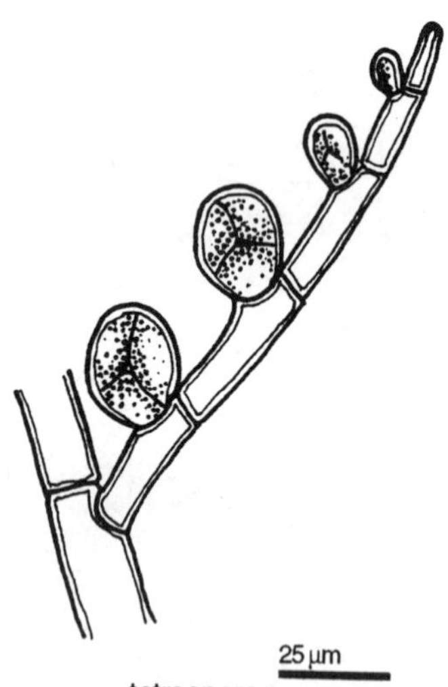

tetraspores

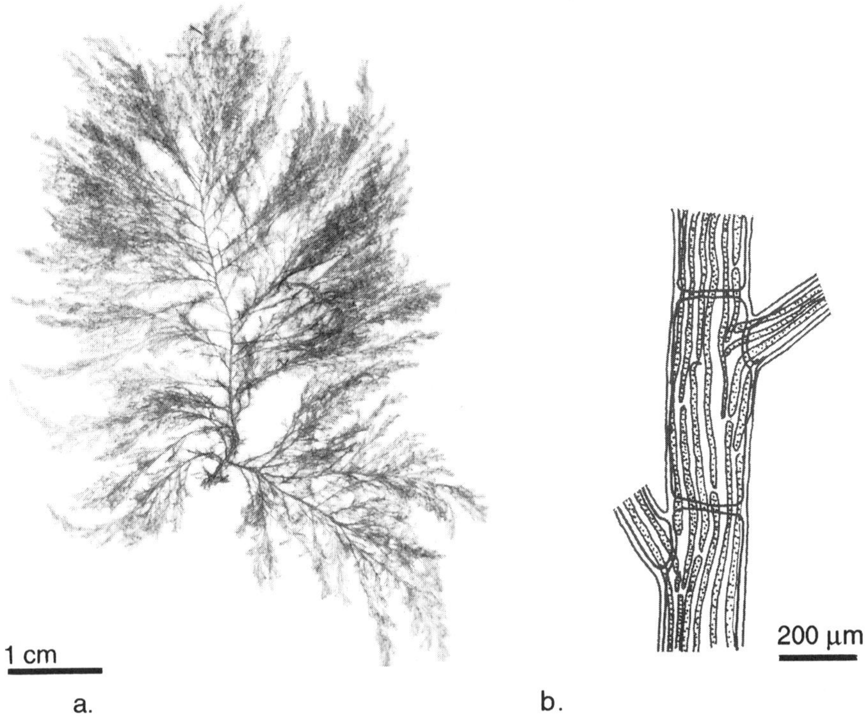

Fig. 84. ***Seirospora interrupta***
a. tuft; b. detail of lower axis with narrow cortical cells

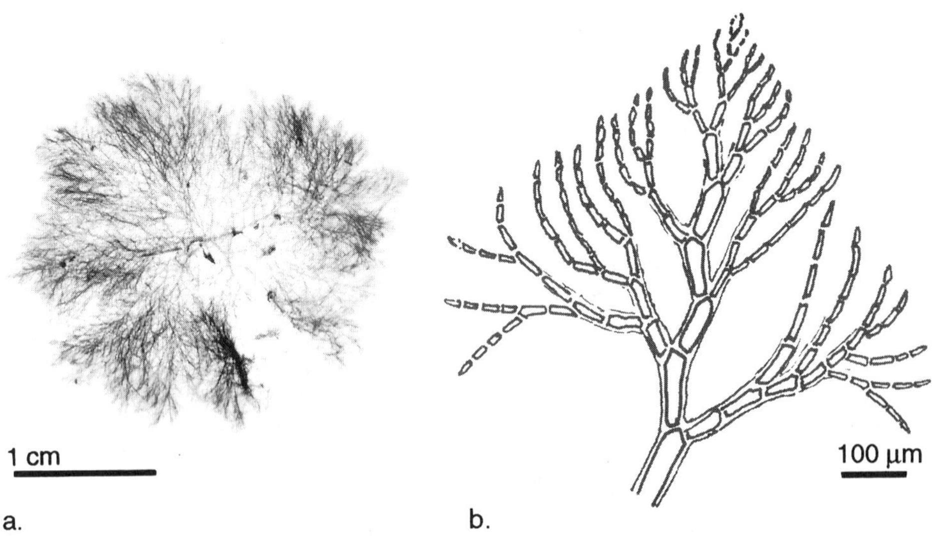

Fig. 85. ***Pleonosporium borreri***
a. tuft; b. detail of branch tip showing triangular pattern

97. **Cells multinucleated; axes corticated at lest near base** 98
97. **Cells uninucleated; cortication if present, very limited; descending, multicellular, rhizoidal filaments often originating from basal cells** 99

98. **Cortication heavy, extending from base to at least the main branches; branching alternately bilateral**; thallus spongy, pyramidal, dark red to brownish, up to 20 cm long, with a distinct main axis, up to 1 mm wide; branches tapering sharply, curved inward and crowded at tips; this is the coarsest of the local *Callithamnion* species; common, found year round, on the open coast and in estuaries, in the lower intertidal and subtidal zones, down to -10 m, epiphytic on eelgrass and other algae, or floating in the drift; also found in tide pools .... *Callithamnion tetragonum* (Fig. 86)

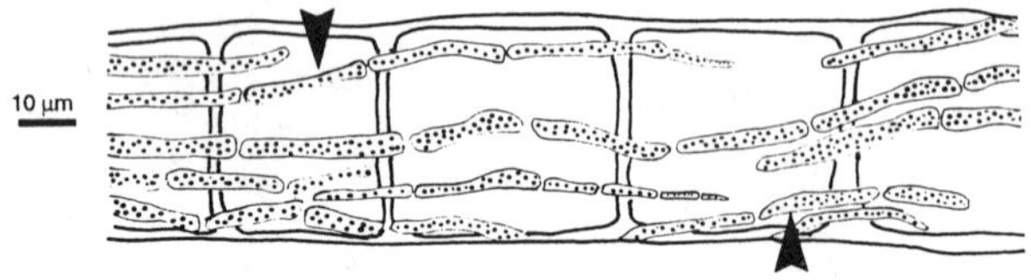

C. tetragonum, cortication on upper axis

98. **Branching subdichotomous, crowded at tips and forming a flat top; irregular or alternate at base, with colorless hairs at tips**; thallus a rounded, delicate tuft, up to 6 cm high, bright pink to brownish red, with a little cortication at base; found year round, but common from early summer through fall, mostly in bays and estuaries, in low tide pools and in the subtidal zone, down to -15 m, epiphytic on eelgrass and other algae, but also attached to rocks .... *Callithamnion corymbosum* (Fig. 87)

99. **Thallus very delicate and soft; without an obvious main axis or basal branching**; entangled; tufts rounded, up to 7 cm high, light pink to red; branching alternate; cells often 5 diameters long; found in the summer, more common from Cape Cod southward, in the shallow subtidal zone, in protected waters of bays and estuaries, epiphytic on eelgrass and other algae; also growing in tide pools .... *Aglaothamnion westbrookiae* * (Fig. 88)
99. **Obvious branching near base; main axis usually visible** (up to 1 mm wide); branching alternately bilateral, free; thallus firm, bushy, with a coarser growth than the 2 previous species; deep pink to purplish red, up to 8 cm high; found year round, from Cape Cod southward, on the open coast and in estuaries, in the subtidal zone, down to -12 m, epiphytic on other algae and on eelgrass, in sheltered areas or sheltered niches .... *Aglaothamnion roseum* (Fig. 89)

* Rueness and L'Hardy-Halos (1991).

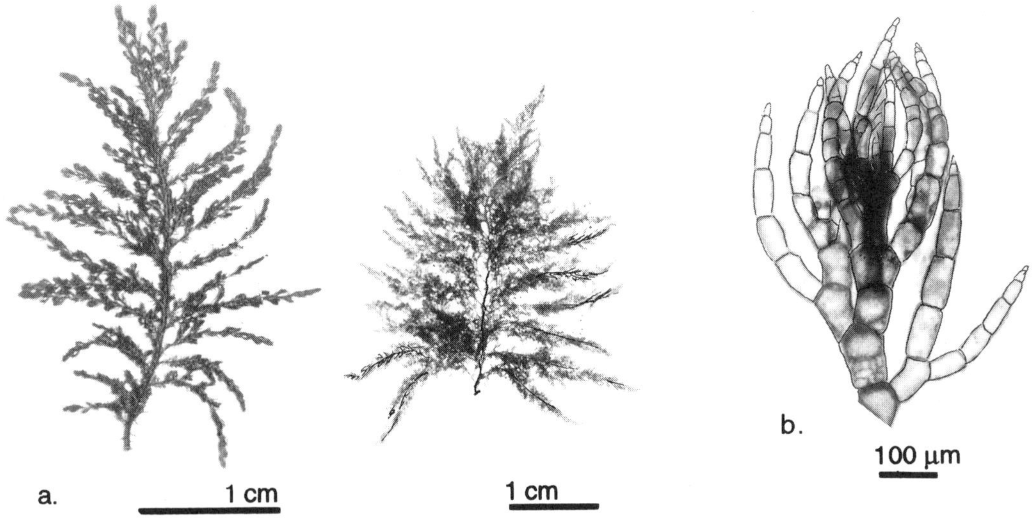

Fig. 86. ***Callithamnion tetragonum***
a. whole thalli showing variation; b. detail of branch tip

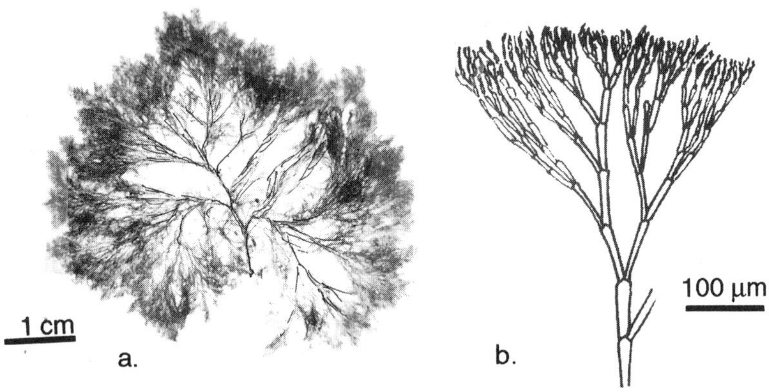

Fig. 87. ***Callithamnion corymbosum***
a. tuft; b. detail of branch with flat top

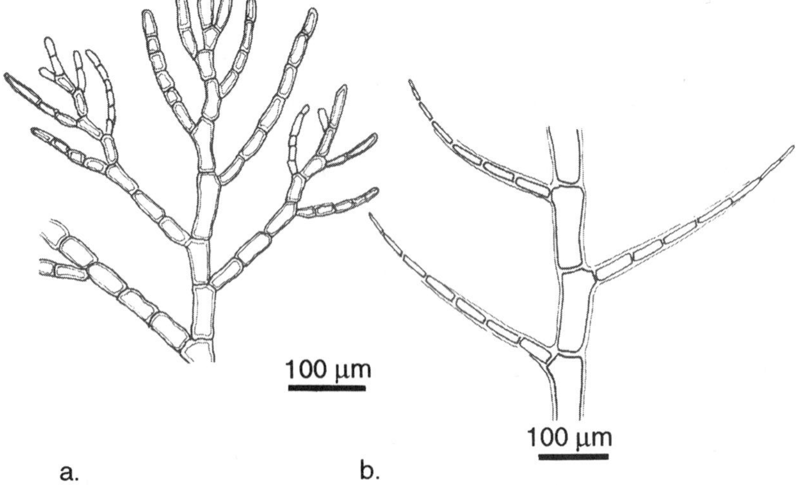

Fig. 88. ***Aglaothamnion westbrookiae***
a. pattern of branching at tip
b. branching pattern in mid section

Fig. 89. ***Aglaothamnion roseum***
tuft

100. **Branches soft, delicate, flat and feather-like; with coarse and cylindrical main axes; tips of branches uniseriate** (see p. 90, step 39, Fig. 37) .... *Plumaria plumosa*
100. **Branches otherwise** 101

101. **Hair-like branches forming tufts with somewhat flattened tops;** thallus up to 1 mm and wiry at base, softer in younger parts, bushy, up to 30 cm tall; axes denuded below; hair-like clusters of branches fall off during the summer; with a typical polysiphonous construction (p. 112, step 81) of 6-7 pericentral cells, obscured by heavy cortication; rosy red near tips, with brownish to black axes; found year round, on the open coast and in estuaries, usually attached to rocks, in low tide pools and in the subtidal zone, down to -15 m; often floating in the drift .... *Rhodomela confervoides* (Fig. 90)

101. **Tufts with somewhat flattened tops absent** 102

102. **Pincers present at tips; branches banded or beaded** (see p. 108, step 76, Figs. 67-71) .... *Ceramium* spp.
102. **Pincers absent** 103

103. **Covered with delicate visible hairs; axes with alternating short and long cross bands** (see p. 106, step 71, Fig. 65) .... *Spyridia filamentosa*
103. **If hairs (trichoblasts) are present, they are only visible with a microscope**; 4-24 pericentral cells (best seen in cross-section) forming cross bands of even sizes, as long as the axial cell; with or without cortication 104

104. **Short, black tufts, epiphytic on *Ascophyllum nodosum*** (see p. 98, step 51, Fig.49) .... *Polysiphonia lanosa*
104. **Not as above** 105

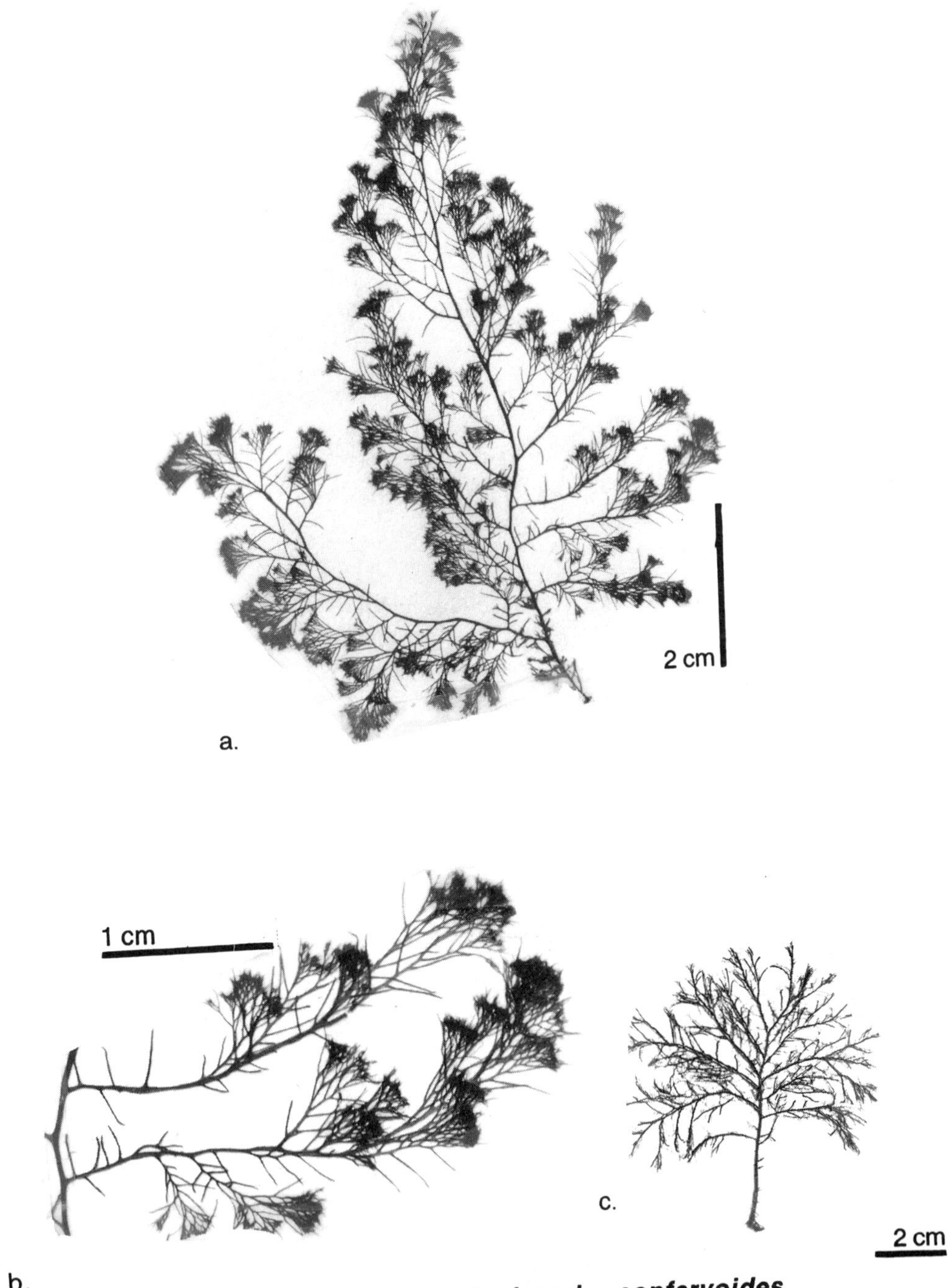

Fig. 90. ***Rhodomela confervoides***
a. spring specimen with tufts of hair-like branches
b. detail of branches with tufted branchlets
c. fall specimen without tufts

105. **Thallus creeping (prostrate, dorsiventral);** tufts brownish purple to almost black; horizontal filaments attached by long rhizoids, and forming upright tufts, up to 3 cm high; main axes with 4-10 pericentral cells, transversally divided into 2 tiers; branches with 4 pericentral cells only and becoming uniseriate at tip; uncommon; found year round, in New Hampshire and to the south, in protected salt marsh flats, growing in brackish waters of low salinity, attached to rocks, shells, wood and mud, or epiphytic on the base of *Spartina* spp.; often associated with *Calloglossa leprieurii* .... *Bostrychia radicans* (Fig. 91)

105. **Thallus upright, not growing flat on substrate** (*Polysiphonia* spp. ♦7 of 10) 106

106. **Four pericentral cells** (see p. 112, step 81) 107

106. **More than 4 pericentral cells** 112

107. **Cortication absent, even at base** 108

107. **Cortication present, at least near base** 109

108. **Trichoblasts (drawing below) rare or absent; found on unprotected shores, in relatively cold waters**; soft, bushy tufts with a creeping base and no distinct axis; bright pink to burgundy, up to 30 cm long; branching sparse at base, more abundant at tips; very common year round, on the open coast and in estuaries, in the mid intertidal and subtidal zones, down to -25 m, attached to rocks, shells and mud, or epiphytic on other algae .... *Polysiphonia stricta* (Fig. 92)

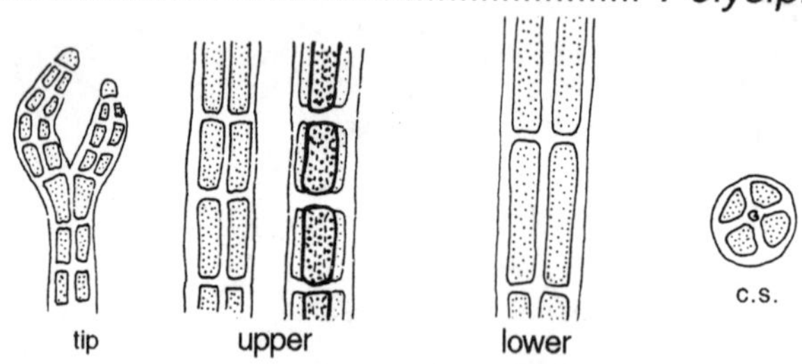

P. stricta

tip of branch, surface views, side view and cross-section

108. **Trichoblasts occasionally present; found in brackish waters, and in warm, protected, shallow and muddy coves;** soft tufts, dark purplish, up to 15 cm long, without a main axis; rhizoids often abundant; found commonly on muddy shores, in salt marshes and estuaries, on wood, at the base of *Spartina alterniflora* and on other substrata; extends up into fresh water .... *Polysiphonia subtilissima*

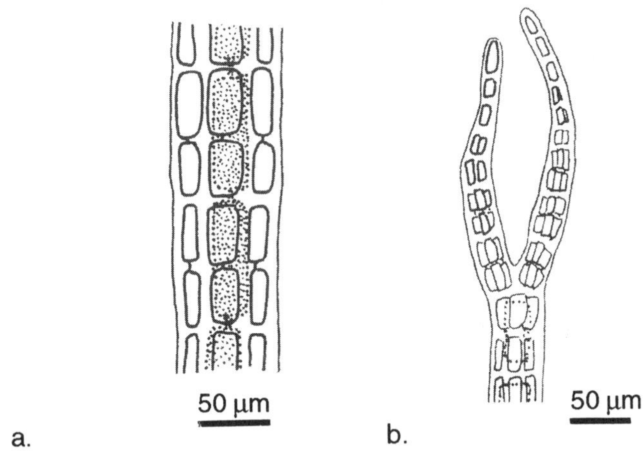

Fig. 91. ***Bostrychia radicans***
a. detail of pericentral cells
b. detail of branch tip with uniseriate end

Fig. 92. ***Polysiphonia stricta***
tuft with matted base

(Fig. 91a and b, redrawn after Schneider and Searles, 1991)

109. **Cortication if present, restricted to base only, very scanty; numerous trichoblasts at tips and below (deciduous below); main axes and branches distinct;** thallus bushy, soft or stiff; pinkish to light reddish brown or blackish, up to 12 (20) cm long; branching irregular near base, alternate above; with many tapering adventitious (side) branches, 1-2 mm long; common year round, on the open coast and in estuaries, in the lower intertidal and subtidal zones, in protected shallow waters, attached to rocks, wood and shells, or epiphytic on eelgrass and other algae; often found floating in the drift .... *Polysiphonia harveyi* (Fig. 93)

109. **Cortication extending at least to main branches** 110

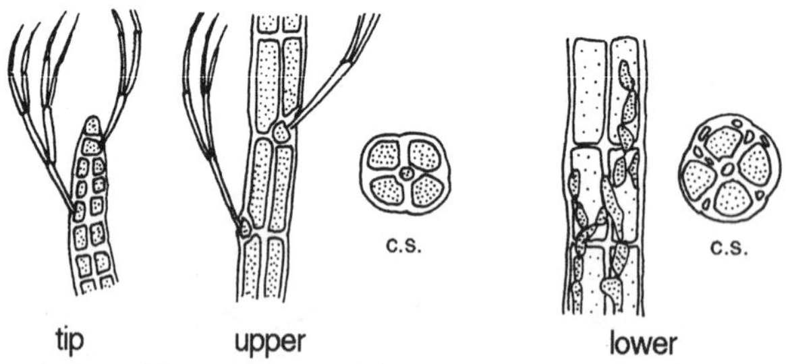

P. harveyi, tips of branches, surface views and cross-sections

110. **Cortication extending to most branches** 111

110. **Cortication restricted to axes and larger branches; thallus very soft and slippery;** tufts delicate, brownish pink to purplish red, up to 40 cm long; axes dividing near base into few, long, primary branches, with rebranching mostly near tips; abundant trichoblasts at tips; found year round, from Long Island Sound northward, on the open coast and in estuaries, in the subtidal zone, down to -15 m, attached to rocks or shells, and epiphytic on other algae .... *Polysiphonia flexicaulis* (Fig. 94)

111. **Cortication limited to main axes and branches; trichoblasts abundant**; thallus bushy, triangular, reddish brown to straw-like, up to 15 cm long; little branching at base, more abundant towards tips; abundant adventitious (side) branching; found year round, from Cape Cod southward, in protected, shallow waters of bays and estuaries, in the subtidal zone, attached to rocks, and epiphytic on other algae .... *Polysiphonia fibrillosa* (Fig. 95)

111. **Completely corticated, except at tips; trichoblasts present;** thallus bushy, somewhat stiff, coarse and light red below, but soft and bright red near tips, up to 30 cm long; abundant tip branches grow in late winter and fall off in summer; found year round, on the open coast and in estuaries, in deep waters, attached to stones and shells .... *Polysiphonia elongata* (Fig. 96)

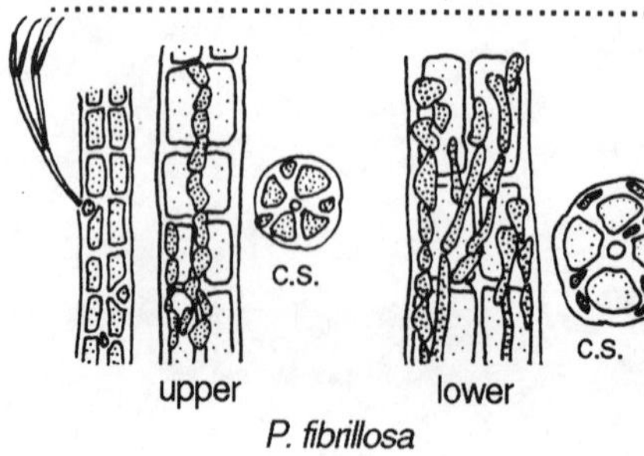

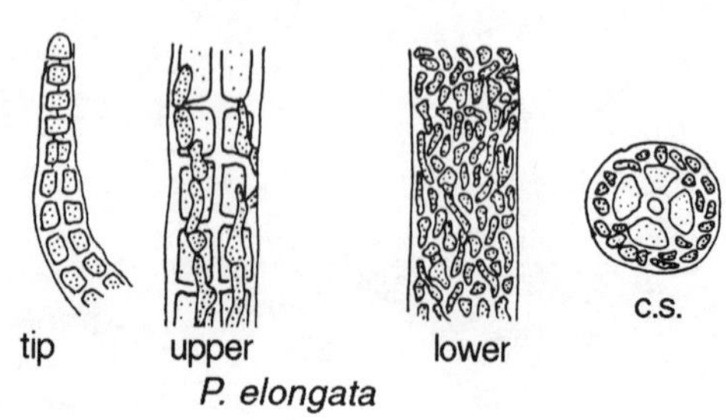

surface views and cross-sections

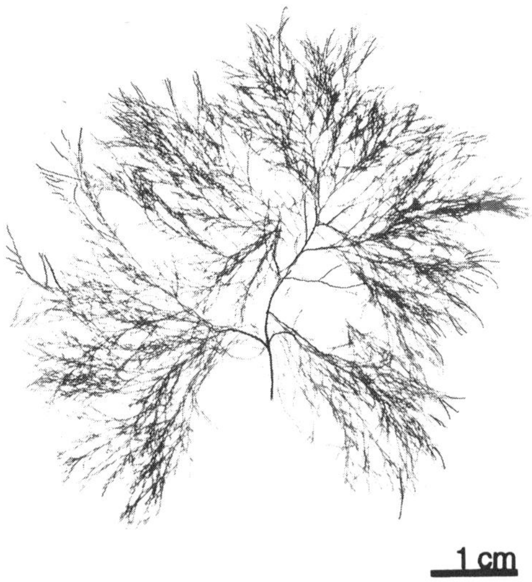

Fig. 93. ***Polysiphonia harveyi***
portion of branch showing typical pattern

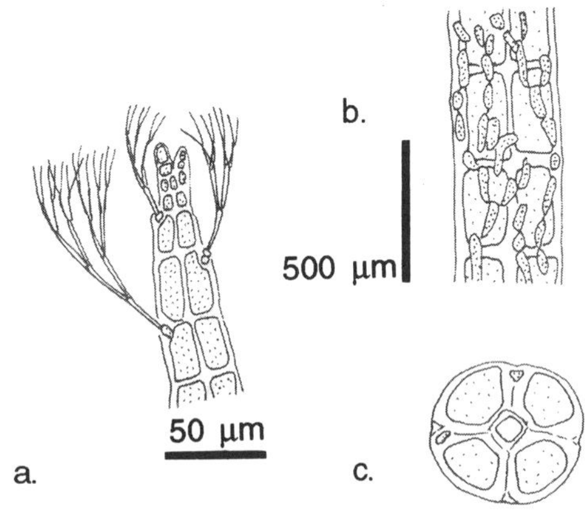

Fig. 94. ***Polysiphonia flexicaulis***
a. tip of branch with trichoblasts
b. cortication on main axis
c. cross section of branch

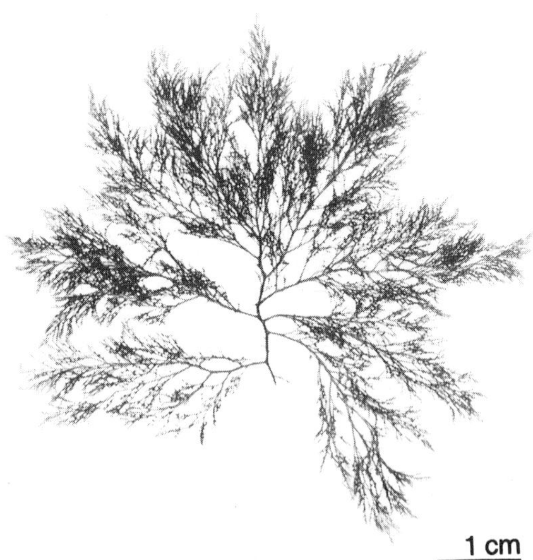

Fig. 95. ***Polysiphonia fibrillosa***
portion of branch showing typical pattern

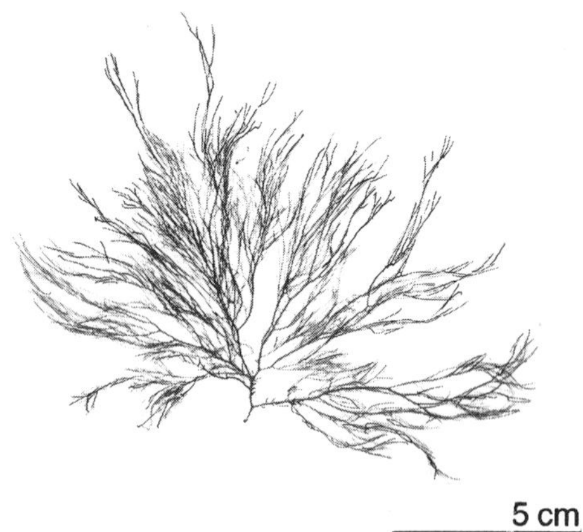

Fig. 96. ***Polysiphonia elongata***
portion of branch showing typical pattern

112. **Six pericentral cells; little cortication; trichoblasts absent**; thallus bushy, up to 25 cm long; somewhat stiff and straw colored below, but very soft and dark purplish red above; very little branching near base; found year round, mostly in the shallow, sheltered waters of bays and estuaries, in the upper subtidal zone, down to -9 m, attached to rocks and wood, or epiphytic on eelgrass and other algae .... *Polysiphonia denudata* (Fig. 97)

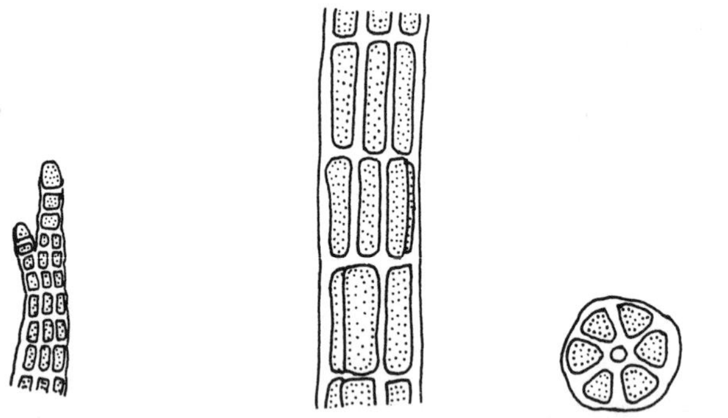

tip of branch   lower axis, surface view   cross-section

P. denudata

112. **More than six pericentral cells** 113

113. **Twelve (8-14) twisted pericentral cells; cortication absent; trichoblasts abundant**; thallus bushy, somewhat firm, brownish or purplish red to black, up to 20 cm long; few branches below; found year round, on the open coast and in estuaries, in the low intertidal and upper subtidal zones, attached to rocks, and occasionally epiphytic on other algae .... *Polysiphonia nigra* (Fig. 98)

113. **Sixteen to twenty pericentral cells, not twisted; corticated at least near base; deciduous trichoblasts**; thallus bushy, firm, dark purplish red to black, up to 30 cm long; distinct main axis, little branched below; brush-like branches near tips; found year round, on the open coast and in estuaries, in the low intertidal and subtidal zones, down to -15 m, attached to rocks and other algae .... *Polysiphonia fucoides* (Fig. 99)

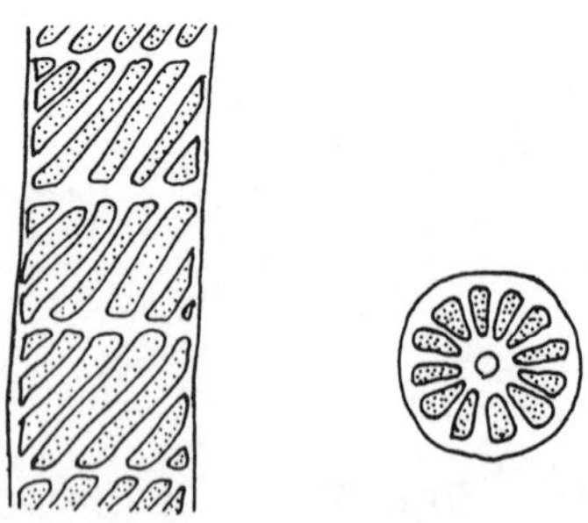

surface view   cross-section

P. nigra

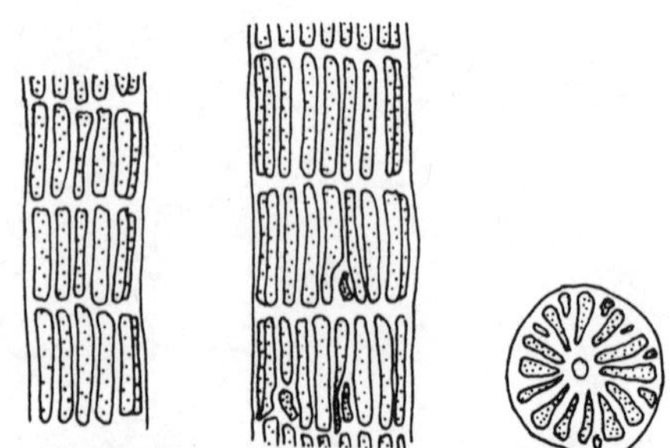

upper axis   lower axis

surface view   cross-section

P. fucoides

Fig. 97. ***Polysiphonia denudata***
portion of branch showing typical pattern

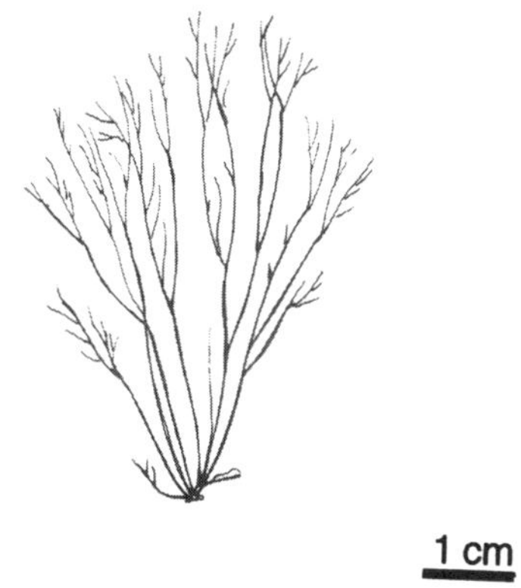

Fig. 98. ***Polysiphonia nigra***
portion of branch showing typical pattern

Fig. 99. ***Polysiphonia fucoides***
portion of branch showing typical pattern

114. **Axes and branches soft and gelatinous** (see p. 100, step 57, Fig. 54) .... *Gloiosiphonia capillaris*
114. **Thallus coarse, with a firm consistency** 115

115. **Axes and branches banded or beaded** (see p. 108, step 76, Fig. 67) . ... *Ceramium nodulosum* (Fig. 100)
115. **No banding or beading visible** 116

116. **Center of cross-section with a solid medulla of round cells only (drawing below); tetraspores cruciate (drawing below); yellowish, greenish, brownish or purplish red, occasionally dark red at base**; large tufts or clumps, up to 35 cm long; branching very irregular, cylindrical or flattened; main axis not evident; often covered with very prominent, dark reproductive lumps; found year round, in the shallow waters of estuaries, in the subtidal zone; extremely variable in shape and color (see p. 94, step 46, Fig. 44, for form with flattened branches) .... *Gracilaria tikvahiae* (Fig. 101)
116. **Medulla filamentous (drawing below); tetraspores zonate (drawing below)** 117

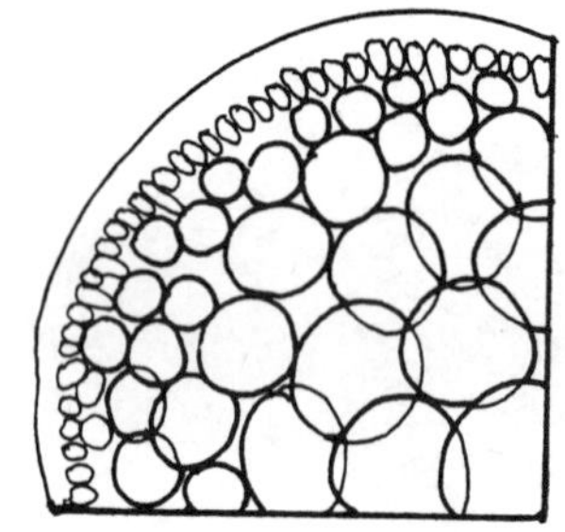

cross-section showing medulla of round cells

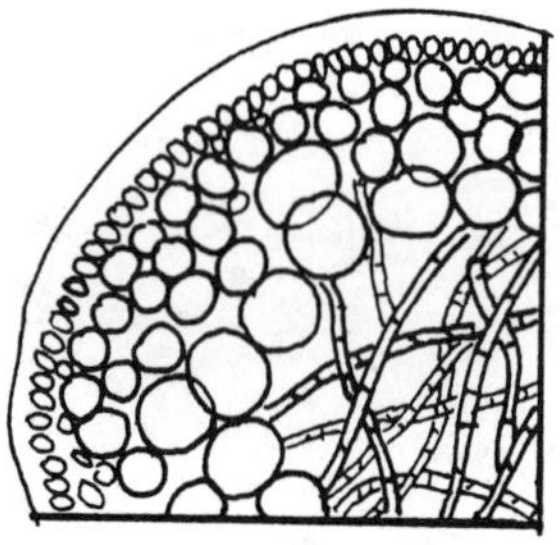

cross-section showing filamentous medulla

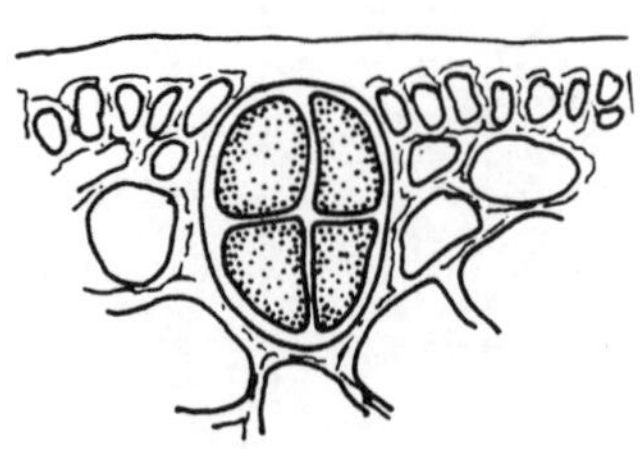

cruciate tetraspores

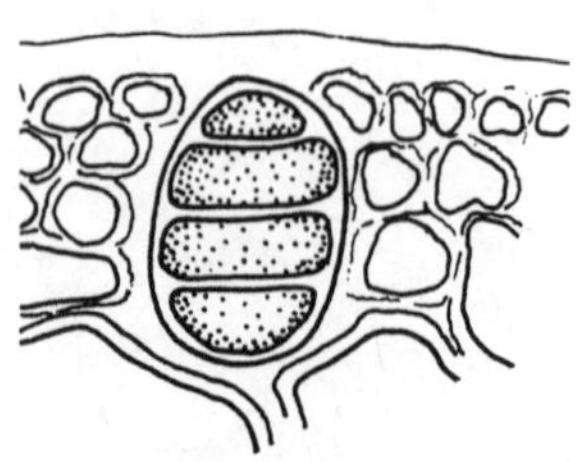

zonate tetraspores

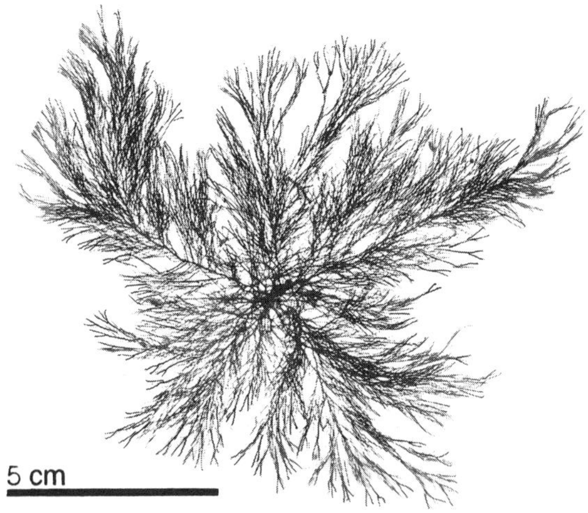

Fig. 100. ***Ceramium nodulosum***
old specimen without pincers

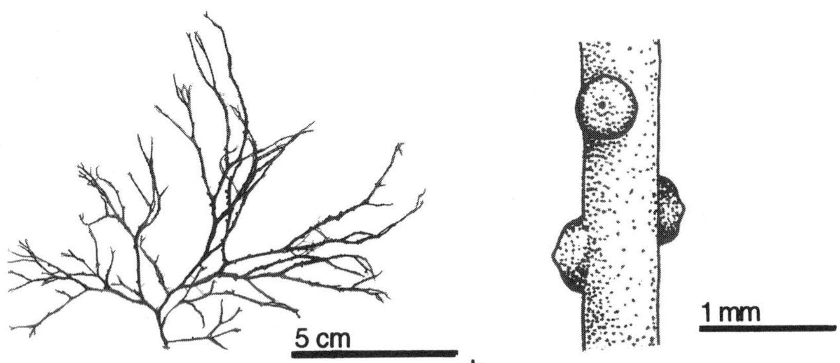

a. b.

Fig. 101. ***Gracilaria tikvahiae***
a. cylindrical form, whole specimen;
b. section of branch with reproductive structures (carposporophytes)

117. **Bright red to burgundy red; axes and branches cylindrical, of equal width** (1-4 mm); thallus firm, fleshy, up to 40 cm long; alternately branched, 1-3 (4) times, branches tapered at both ends; dark red reproductive lumps often present; common year round, mostly from Cape Cod southward, but also recorded occasionally up to New Hampshire; growing mostly in estuaries, in the low intertidal and subtidal zones, attached to stones and shells ....*Agardhiella subulata* (Fig. 102)

117. **Distinct main axis, 1-3 mm wide, wider than branches; single apical cell at tip of young branches**; thallus bushy, firm, fleshy, up to 50 cm long; branches abundant, 3-5 times irregularly branched; smaller branches tapered at both ends; light rosy red to deep purplish red, also yellowish or greenish; reproductive lumps often visible as a slight swelling; very common year round, on the open coast and in estuaries, from the mid-intertidal and subtidal zones, down to -15 m, attached to rocks or shells, and occasionally epiphytic on other algae; sometimes growing in large populations; extremely variable (see p. 108, step 74, Fig. 66, for variety with tendrils) .... *Cystoclonium purpureum* (Fig. 103)

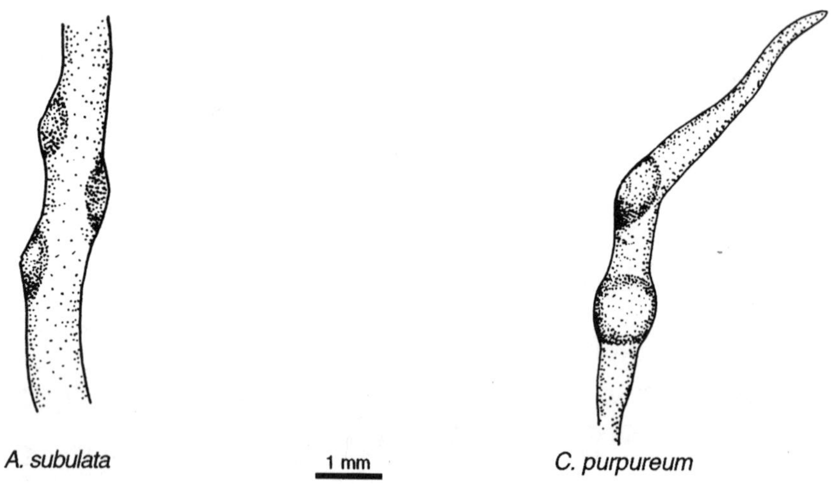

female reproductive structures (swellings are carposporophytes)

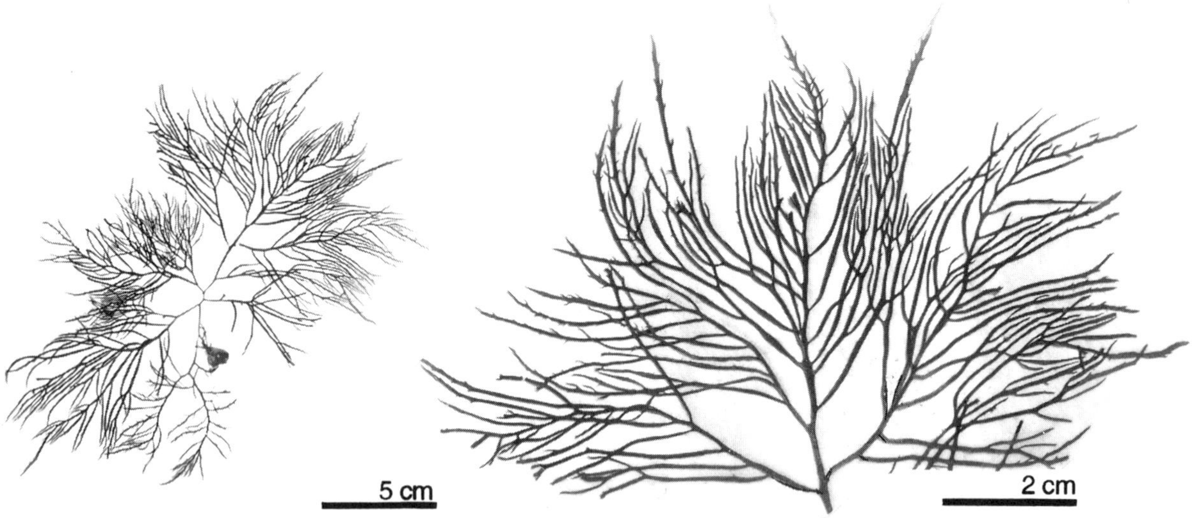

2 cm

a. b.

Fig. 102. ***Agardhiella subulata***
a. whole thallus; b. detail of branching pattern

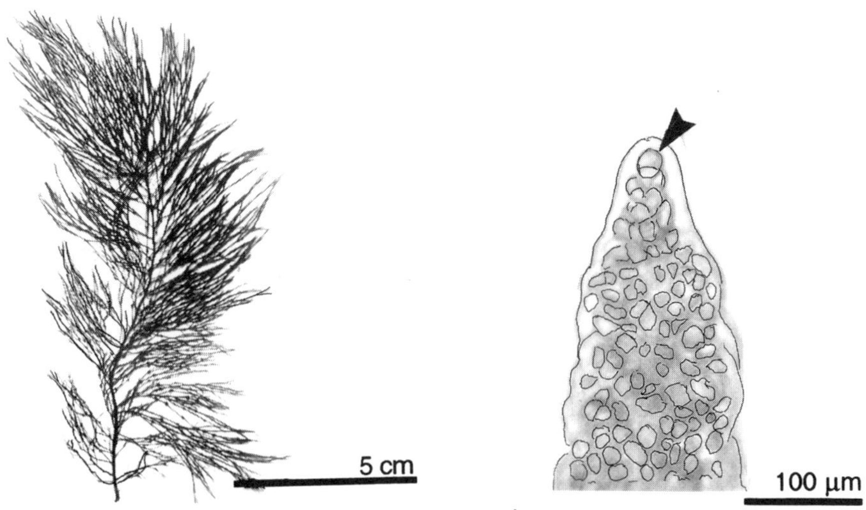

a. b.

Fig. 103. ***Cystoclonium purpureum***
a. thallus without tendrils; b. branch tip with single apical cell

GLOSSARY

Page and step numbers
refer to diagrams

Adventitious: Usually referring to branches originating from an unusual place or in an irregular or sporadic pattern; produced secondarily.

Alga (pl. **algae**): Common term used for a group of generally photosynthetic, eukaryotic organisms, without complex tissues or multicellular sex organs.

Alternate branching: Type of branching in which the successive branches arising from different levels, are on opposite sides of the axis (p. 123, Fig. 88).

Antheridium (pl. **antheridia**): The male reproductive structure that produces motile male gametes (p. 60, step 44).

Apical: At, or near the tip.

Axis: Main filament or section which bears branches.

Bilateral: Usually referring to branches or reproductive structures located on both sides of an axis.

Bisporangium (pl. **bisporangia**): Sporangium which produces two spores (bispores).

Biseriate: Cells arranged in two rows (p. 14, step 4).

Blade: Flattened, leaf-like portion of an alga.

Branchlet: The smallest, pigmented division of a branched organism (p. 102, step 64).

Calcareous, calcified: Incrusted with lime, including calcium and magnesium carbonates, and stone-like.

Carposporophyte: A complex, minute generation of the Rhodophyta, which remains attached to the female gametophyte and produces carpospores; often visible with the naked eye as a rounded bump.

Conceptacle: Small, sunken cavity with its opening at the surface of the thallus; produces reproductive cells (p. 60, step 44; p. 76, step 10).

Crust: Body of organism growing flat on substrate, either smooth or rough, hard or soft, sometimes brittle.

Coralline: Algae incrusted with lime, very hard and coral-like.

Cortex: Group of cells, generally photosynthetic and somewhat compact, lying between the outer most layer of cells and the medulla (p. 58, step 40).

Cortication: A secondary layer of small, surface cells produced by the supporting axial cell (p. 108, step 75).

Cruciate: In the shape of a cross; generally applied to a tetraspore arrangement produced by two successive divisions, at right angle to each other (p. 132, step 116), or to a branching pattern (p. 117, Fig. 78b).

Crust, crustose: A thin algal structure which grows flat against the substrate.

Dichotomous branching: Y shaped, even and regular branching or forking.

Ecad.: A non-heritable growth form produced by specific ecologial conditions.

Endophytic: Growing within another plant or alga (not necessarily parasitic).

Endozoic: Growing within an animal.

Epiphytic: Growing on a plant or an alga (not necessarily parasitic).

Epizoic: Growing on an animal.

Fascicle: Short, broom-like cluster of branches.

Fertile: Bearing reproductive structures.

Filament: Chain of cells, joined end to end and forming a hair-like strand.

Fucoid: Group of algae belonging to the Fucales order, and resembling *Fucus*.

Gametophyte: Phase of a life cycle which produces gametes.

Gametangium (pl. **gametangia**): Cell which produces gametes.

Gland cell: A specialized, distinct, rounded and refractive (glassy) cell, with possible roles in storage and secretion, found in the Rhodophyta (p. 116, step 89; p. 118, step 94).

Holdfast: A root-like, single cell or group of cells which anchor the alga on to the substrate.

Internode: Surface of a cell between the two end-walls in a filament, or portion of an axis between branching points.

Intertidal zone: Area of the coast line located between the highest high tide and the lowest low tide.

Macroscopic: Large enough to be seen with the naked eye.

Medulla: Group of cells, usually colorless, located on the inside of the cortex, either loosely or compactly organized (p. 58, step 40; p. 132, step 116).

Micrometer (µm): 1/1000 of a millimeter.

Microscopic: Too small to be seen with the naked eye; magnification is required for observation.

Midrib: A slightly raised, longitudinal strip or line running through the central axis of a blade (p. 62, step 46; pp. 84-87, Figs. 24-30).

Monosporangium (pl. **monosporangia**): Sporangium which produces a single spore (monospore) (p. 113, Figs. 72).

Multiseriate: With more than two rows of cells.

Nemathecium (pl. **nemathecia**): A reproductive structure of the Rhodophyta consisting of a raised surface, supporting reproductive cells; usually visible with the naked eye as a small lump.

Node: Area of connection between two cells in an algal filament, or point from which a branch develops

Oogonium: A single-celled reproductive structure which produces eggs (p. 60, step 44).

Opposite branching: Type of branching in which two branches originating from one node, are at a 180° from each other (p. 117, Fig. 77 and 79).

Paraphysis: Sterile, hair-like filament located next to reproductive structures.

Pericentral cells: Cells produced by, and surrounding the axial cells, and as long as the axial cell; found in the Rhodophyta (p. 112, step 81).

Pericyst: Specialised cell, more transparent than others in living specimens, but becoming darker in dried specimens (p. 45, Fig. 14).

Pit-connection: Connection between two adjacent cells, usually plugged; visible, particularly in the Rhodophyta, as an hour-glass strand.

Plastid: Organelle within a cell, specialized for a certain function and variously pigmented, *e.g.* chloroplast (p. 40, step 7; p. 46, step 20).

Plurilocular (sporangium or gametangium): Reproductive structure of the brown algae, usually elongated and with many small chambers; produced by mitosis (p. 49, Figs. 20b, 21b, and 22b).

Pluriseriate: With more than two rows of cells.

Polysiphonous: With tiers of pericentral cells surrounding an axis; each tier covering a corresponding axial cell, and equal to it in length; found in the Rhodophyta (p. 112, step 81).

Polysporangium (pl. **polysporangia**): Sporangium which produces more than four spores (polyspores) (p. 120, step 96).

Pyrenoid: Roundish body in the chloroplast of the green algae associated with starch reserves; also found in other phyla but not involved with starch (p. 26, step 27 and 28).

Receptacle: Inflated portion of a branch or branch tip bearing sunken reproductive cavities (conceptacles) in the Fucales (p. 60, step 44).

Recurved: Bent backwards.

Rhizoid: Outgrowth of the basal part of some algae, made up of one cell, or a chain of cells, growing downward or horizontally, usually serving for attachment and/or absorption.

Seirosporangium (pl. **seirosporangia**): Sporangia in linear series, usually branched and produced at the tip of a branch; produces seirospores (p. 120, step 95).

Septate: With transversal cell walls forming distinct cellular structures.

Sorus (pl. **sori**): Cluster of reproductive stuctures.

Sporangium (pl. **sporangia**): Reproductive cell which bears the spores.

Sporophyll: Leaf which bears spores; in the algae restricted to some Phaeophyta (p. 55, Fig. 33).

Sporophyte: Phase of a life cycle which produces spores.

Stipe: Stem-like portion of thallus; cylindrical or flattened.

Subdichotomous branching: Branching which appears dichotomous.

Subtrate (pl. **substrata**): Surface on which an organism is attached.

Subtidal: Area of the coast line located below the lowest tide level and always submerged.

Tetrahedral: Arrangement of tetraspores found in the Rhodophyta, in which the spores are produced by meiosis and are pyramidal; only three spores are visible at one time (p. 120, step 96).

Tetrasporangium: Sporangium which produces four spores (tetraspores), usually by meiosis (p. 76, step 10; p. 120, step 96; p. 132, step 116).

Thallus (pl. **thalli**): Simple, plant-like body, without differentiation into organs (roots, stems and leaves), or complex tissues (in particular, vascular bundles).

Trichoblast: Unpigmented, branched or branched hair-like filament produced at the edge, surface or tip of an organism, in the Rhodophyta (p. 128, steps 109-112).

Unilocular sporangium: Reproductive structure of the brown algae in which spores are produced in a single chamber, by meiosis; usually rounded or somewhat oval (p. 49, Figs. 20c, 21c, 22c, and 23).

Uniseriate: Cells arranged in one row (p. 14, step 4; p. 46, step 19; p. 112, step 81).

Vegetative: Not reproductive by gametes or spores.

Vesicle or air bladder: Rounded, bladder-like structure, usually filled with gases (p. 61, Fig 43; p. 62, step 46).

Zonate: Arrangement of tetraspores in which the spores are arranged in a linear series, with parallel walls (p. 132, step 116).

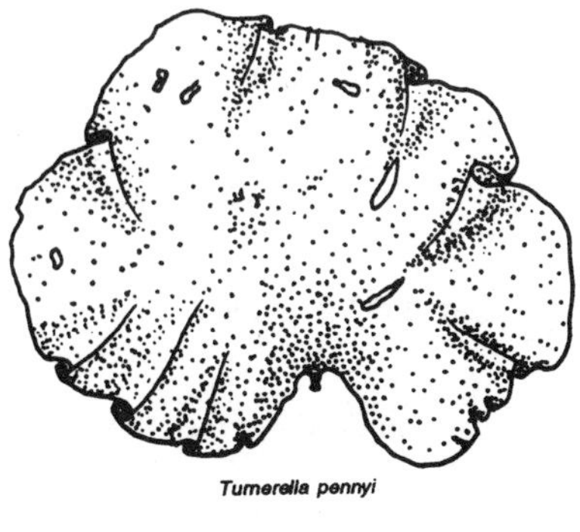

Turnerella pennyi

BIBLIOGRAPHY

Adey, W. H. and Adey, P. J, 1973. Studies of the biosystematics and ecology of the epilithic crustose corallinaceae of the British Isles. Br. phycol. J. 8: 343-407.

Bird, C. J. and J. L. McLachlan. 1992. Seaweed Flora of the Maritimes 1. Rhodophyta - The Red Algae. Biopress Ltd., Bristol, England. 176 pp.

Blair, S. M. 1983. Taxonomic treatment of the *Chaetomorpha* and *Rhizoclonium* species (Cladophorales; Chlorophyta) in New England. Rhodora 85: 175-211.

Blair, S. M., A. C. Mathieson and D. P. Cheney. 1982. Morphological and electrophoretic investigations of selected species of *Chaetomorpha* (Chlorophyta; Cladophorales). Phycologia 21: 164-172.

Bold, H. C. and M. J. Wynne.1985. Introduction to the Algae. 2nd Ed., Prentice Hall. Englewood Cliffs, N.J. 720 pp.

Brummit R. K. and C. E. Powell. 1992. Authors of Plant Names. Royal Botanic Gardens, Kew. 732 pp.

Burrows, E. M. 1991. Seaweeds of the British Isles. Volume 2. Chlorophyta. Natural History Museum, London. 238 pp.

Cardinal, A. 1964. Étude sur les Ectocarpacées de la Manche. Nova Hedwgia 15. 170 pp.

Dawson, E. Y. 1956. How to Know the Seaweeds. William C. Brown Co., Dubuque, Iowa. 197 pp.

Dixon, P. S. and L. M. Irvine. 1977. Seaweeds of the British Isles. Rhodophyta. 1. Introduction, Nemaliales, Gigartinales. Natural History Museum, London. 252 pp.

Fletcher, R. L.1987. Seaweeds of the British Isles. III. Fucophyceae (Phaeophyceae).1. Natural History Museum, London. 359 pp.

Hehre, E. J. 197?. A photographic guide to the common genera of marine algae of eastern Long Island, New York. Unpublished manuscript.

Hehre, E. J. 1993. Key to the Chlorophyta, Phaeophyta and Rhodophyta of the southern Gulf of Maine. Unpublished manuscript.

Hehre, E. J. and A. C. Mathieson. 1994. Algal nomenclature for New Hampshire seaweeds. Unpublished manuscript.

Hillson, C. J. 1977. Seaweeds. A Color Coded Illustrated Guide to the Common Marine Plants of the East Coast of the United States. Keystone Books, Penn. State Univ. Press, Univ. Park. 194 pp.

Hine, A. E. 1977. A Glossary of Phycological Terms for Students of Marine Macroalgae. St. Alden's in the weeds, Miami, Fl. 91 pp.

Irvine, L. M. 1983. Seaweeds of the Brithish Isles. I. Rhodopyta. 2A. Cryptonemiales (sensu stricto), Palmariales, Rhodymeniales. Natural History Museum, London. 115 pp.

Irvine, L. M. and Y. M. Chamberlain. 1994. Seaweeds of the British Isles. Volume 1. Rhodophyta. Part 2B. Corallinales, Hildenbrandiales. Natural History Museum, London. 276 pp.

Jónsson, S. 1962. Recherches sur des Cladophoracées marines. Structure, reproduction, cycles comparés, conséquences systématiques. Ann. Sci. Nat. Bot. Ser. 12, 5: 25-264.

Kingsbury, J. M. 1969. The Seaweeds of Cape Cod and the Islands. The Chatham Press Inc., Chatham, Mass. 212 pp.

Lee, T. F. 1986. The Seaweed Handbook. An Illustrated Guide to the Seaweeds from North Carolina to the Arctic. 2nd Ed. Dover Publ., New York. 216 pp.

Maggs, C. A. and M. H. Hommersand. 1993. Seaweeds of the British Isles. Volume 1. Rhodophyta. Part 3A Ceramiales. Natural History Museum, London. 444pp.

Mathieson, A. C. and E. J. Hehre. 1986. A synopsis of New Hampshire seaweeds. Rhodora 88: 1-65.

Prud'homme van Reine, W. F. 1982. A Taxonomic revision of the European Sphacelariaceae (Sphacelariales, Phaeophyceae). Leiden Univ. Press, Bot. Ser. 6: 1-293.

Rice, E. L. and A. R. O. Chapman. 1985. A numerical taxonomic study of *Fucus distichus* (Phaeophyta). J. Mar. Biol. Ass. U.K. 65:433-459.

Rueness, J. and M. T. L'Hardy-Halos. 1991. *Aglaothamnion westbrookiae* sp. nov. (Rhodophyta), a species previously confused under the name *Callithamnion byssoides*. J. Phycol. 27: 649-652.

Schneider, C. W. and R. B. Searles. 1991. Seaweeds of the Southeastern United States, Cape Hatteras to Cape Canaveral. Duke Univ. Press. Durham, N.C. 553 pp.

Schneider, C. W., M. M. Suyemoto and C. Yarish. 1979. An Annotated Checklist of Connecticut Seaweeds. State Geological and Natural History Survey of Connecticut, Bull. 108. 20 pp.

Sears, J. R. 1977. Key to the common marine genera of Chlorophyta, Phaeophyta and Rhodophyta of southern Massachusetts. Unpublished manuscript.

Sears, J. R. and R. T. Wilce. 1973. Sublittoral benthic marine algae of southern Cape Cod and adjacent islands: *Pseudolithoderma paradoxum* sp. nov. (Ralfsiaceae, Ectocarpales). Phycologia 12: 75-82.

Sears, J. R. and R. T. Wilce. 1975. Sublittoral benthic marine algae of southern Cape Cod and adjacent islands: seasonal periodicity, associations, diversity, and floristic composition. Ecol. Monogr. 45: 337-365.

Sears, J. R. *et al.* In preparation. Key to the Benthic Marine Algae of the Northeastern Coast of North America. NEAS.

Silva, M. R. W. N. de and E. M. Burrows. 1973. An experimental assessment of the status of the species *Enteromorpha intestinalis* (L.) Link and *Enteromorpha compressa* (L.) Grev. J. Mar. Biol. Ass. U. K. 53: 895-904.

South, G. R. and I. Tittley. 1986. A Checklist and Distributional Index of the Benthic Marine Algae of the North Atlantic Ocean. Huntsman Marine Laboratory and British Museum (Nat. Hist), St. Andrews and London. 76 pp.

Taylor, W. R. 1957. Marine Algae of the Northeastern Coast of North America. 2nd Ed. Univ. Michigan Press, Ann Arbor, Mi. 509 pp.

Taylor, S. L., and M. Villalard-Bohnsack. 1985. Seaweeds of the Connecticut Shores. 3rd Ed. Conn. Arboretum, Bull. no. 18, New London, Conn. 36 pp.

Villalard-Bohnsack, M. and M. M. Harlin. 1997. The appearance of *Grateloupia doryphora* (Halymeniaceae, Rhodophyta) on the northeast coast of North America. Phycologia (in press).

Villalard-Bohnsack, M., M. P. Peckol and M. M. Harlin. 1988. Marine macroalgae of Narragansett Bay and Adjacent Sounds. In (R. G. Sheath and M. M. Harlin, eds) Freshwater and Marine Plants of Rhode Island. Kendall/Hunt Publishing Company, Dubuque, Iowa. 149 pp.

Wilkinson, M. and E. M. Burrows. 1972. An experimental taxonomic study of the algae confused under the name *Gomontia polyrhiza*. J. Mar. Biol. Ass. U. K. 52: 49-57.

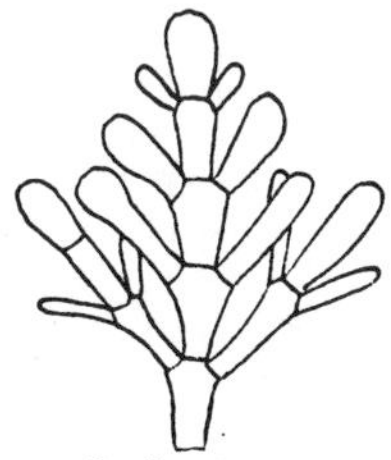

Corallina officinalis

INDEX

When species are listed on more than one page, the pages numbers are given, and the page with the descripton is underlined

Fimbrifolium dichotomum ...

ADDENDUM

Grateloupia doryphora (Montagne) Howe (Halymeniaceae, Rhodophyta), a large, foliose, polymorphic red alga, has recently been recorded for the first time on the northeastern coast of North America. This invasive species is common to the Pacific Ocean, the Mediterranean Sea and the Atlantic Ocean from the British Isles to Angola and from Florida to Uruguay. Since July 1996, specimens have also been observed in large numbers in the lowermost portions of Narragansett Bay, Rhode Island, USA. Reports of new sightings in New England would be greatly appreciated by the author.

Description*

Thalli (Fig. 104a) are foliose, pinkish to brownish red, with a gelatinous and distinctly slippery or silky texture, growing singly, or more often in clumps of up to eight individuals. Blade shape and size range from long and narrow (< 75 cm x < 15 cm) with a tapered base, to broad and short (< 15 cm x < 20 cm) with a somewhat heart-shaped base. Specimens are simple or divided longitudinally or irregularly. Margins are frequently undulated and often bear thread-like or lance-shaped proliferations, sometimes single but more often densely clustered. Margins and tips are frequently lacerated and surfaces often bear signs of grazing. Stipes are short (< 1.5 cm), cylindrical (< 2 mm wide) and occasionally branched. They arise from a small, thin, discoid holdfast (< 5 mm wide).

Blades are 150-600 µm thick and cross-sections (Fig. 104b) show two distinct layer types : the cortex consisting of 4-7 rows of tightly packed, rounded, pigmented cells and the central medulla consisting of unpigmented, loosely interwoven branched filaments joined end to end.

Thalli are attached in the lower intertidal (up to +0.5 m) and in the subtidal zone (down to -1 m), as well as in tide pools throughout the intertidal zone. Substrata include stable bedrock outcrops, loose stones and mussels, in both protected and exposed areas.

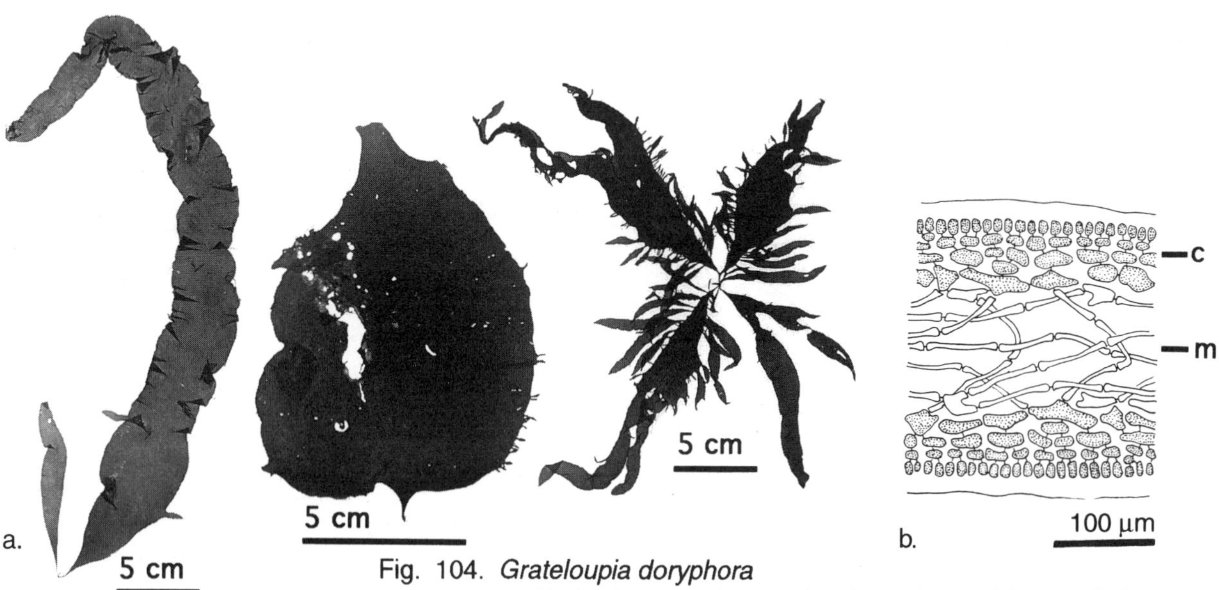

Fig. 104. *Grateloupia doryphora*
a. Herbarium specimens showing polymorphic morphology
b. Cross section with cortex (c) and medulla (m)

* Using the Rhodophyta Key, this species would key out to step 37, p. 90. *Grateloupia doryphora* may be confused with *Palmaria palmata* but the filamentous nature of its medulla clearly separates the two taxa.